Radio and Antenna Days of the Indian Ocean

(RADIO 2012)

IOP Conference Series: Materials Science and Engineering
Volume 44

Flic-en-Flac, Mauritius
24 - 27 September 2012

Editors:

Vikass Monebhurrun **Dominique Lesselier**

ISBN: 978-1-63439-855-8
ISSN: 1757-8981

Printed from e-media with permission by:

Curran Associates, Inc.
57 Morehouse Lane
Red Hook, NY 12571

Some format issues inherent in the e-media version may also appear in this print version.

Printed by Curran Associates, Inc. (2015)

For permission requests, please contact the Institute of Physics
at the address below.

Institute of Physics
Dirac House, Temple Back
Bristol BS1 6BE UK

Phone: 44 1 17 929 7481
Fax: 44 1 17 920 0979

techtracking@iop.org

Additional copies of this publication are available from:

Curran Associates, Inc.
57 Morehouse Lane
Red Hook, NY 12571 USA
Phone: 845-758-0400
Fax: 845-758-2634
Email: curran@proceedings.com
Web: www.proceedings.com

Table of contents

Volume 44

Radio and Antenna Days of the Indian Ocean (RADIO 2012)
24–27 September 2012, Mauritius

Accepted papers received: 5 February 2013
Published online: 23 April 2013

Papers

FOREWORD

It was an honor and a great pleasure for all those involved in its organization to welcome the participants to the 'Radio and Antenna Days of the Indian Ocean' (RADIO 2012) international conference that was held from 24th to 27th September 2012 at the Sugar Beach Resort, Wolmar, Flic-en-Flac, Mauritius.

RADIO 2012 is the first of a series of conferences that is to be regularly organized in the Indian Ocean region. The aim is to discuss recent developments, theories and practical applications covering the whole scope of radio-frequency engineering, including radio waves, antennas, propagation, and electromagnetic compatibility. Following discussions with engineers and scientists from the countries of the Indian Ocean as well as from other parts of the world, a need was felt for the organization of such an international event in this region. The Island of Mauritius, worldwide known for its white sandy beaches and pleasant tropical atmosphere, provided an excellent environment for the organization of the 1st RADIO international conference.

The Local Organizing Committee consisted of scientists from SUPELEC, the University of Mauritius, and the University of Technology, Mauritius. Various members of staff of the University of Mauritius provided help for the organization of the conference. The International Union of Radio Science (URSI) made available technical and financial sponsorship for partial support of young scientists. A number of companies also supported RADIO 2012 ('Platinum': GSMA, ICTA & MMF, 'Gold': CST & FEKO). The event itself was organized in a premier hotel on Mauritius. In this foreword, we would like to take the opportunity again to thank all the people, institutions and companies that made the event such a success.

More than 120 abstracts were submitted to the conference and were peer-reviewed by an international scientific committee. RADIO 2012 overall featured six oral sessions, one poster session and two workshops. Three internationally recognized scientists delivered keynote speeches during the conference. The two workshops consisted of both regular and invited papers.

A call to extended contributions for publication as a volume in the *IOP Conference Series: Materials Science and Engineering* (MSE) both on-line and in print, was made. Following the call, we received thirty-three full papers and all submitted contributions were then peer-reviewed by the co-editors with considerable help from the editorial board. Following this procedure, twenty-seven of them were accepted for publication in this volume.

We believe that this volume will be both an excellent source of pertinant and long-lasting scientific material in the fast evolving fields that are covered by RADIO 2012 as well as good proof of its relevance to the scientific Indian Ocean community and far beyond.

IOP Conf. Series: Materials Science and Engineering **44** (2013) 012001 doi:10.1088/1757-899X/44/1/012001

Automotive Hexband Antenna for AM/FM/GPS/SDARS and AMPS/PCS1900 Cell Phone in an only 65 mm high Housing

J Kammerer, L Reiter and S Lindenmeier

Institute of High Frequency Technology, University of the Bundeswehr Munich, D-85577 Neubiberg, Germany

E-mail: stefan.lindenmeier@unibw.de

Abstract. Nowadays cars are equipped frequently with typical 400 mm long active AM/FM antennas mounted on top close to the rear windshield. In [1], we presented a novel capacitive coupled helical antenna with a height of only 140 mm which performs equivalent to such an 400 mm long antenna. In the next step the antenna height has been reduced to only 56 mm in order to be placed in a low 65 mm housing in combination with other decoupled antennas. The measured results for AM/FM are close to the performance of a whip antenna of 900 mm length. The GPS and SDARS antennas are realized in a combination of two table-formed ring structures with a maximum gain for LHCP at 2339 MHz with 2.9 dBi in zenith and with 5.2 dBi in zenith for RHCP at 1575 MHz with GPS. The VSWR of the cell phone antenna is below 3 for AMPS and PCS1900.

1. Introduction

For 60 years cars have been equipped with rod antennas with a typical length of 800 to 1000 mm usually mounted on a fender of a car for AM/FM reception. At the beginning of the 1970s they were replaced more and more by transistorized active 400 mm long rod antennas mounted on top of a car near the rear windshield. Since 1984 active AM/FM antennas have been integrated inconspicuously in the windshield using extremely thin wires in the laminated safety glass or in the backlite using the heating structure as antenna element. With increasing demand for cell phone antennas at the beginning of the 1990s, compact antennas were installed in small housings on top of the car near the rear window. Antennas for satellite radio reception in the S-band as well as antennas for GPS navigation were implemented frequently in such housings with a height below 65 mm. For some years active helical antennas have been on the market with a height of 200 to 300 mm. In [1], we presented a novel capacitive coupled helical AM/FM antenna with a height of only 140 mm. But it is very tremendously challenging to implement an AM/FM antenna in a housing of only 65mm.

2. Automotive Hexband antenna design principle

In figure 1 the new hexband antenna design principle is displayed. As AM/FM antenna element a capacitive coupled helix antenna is applied

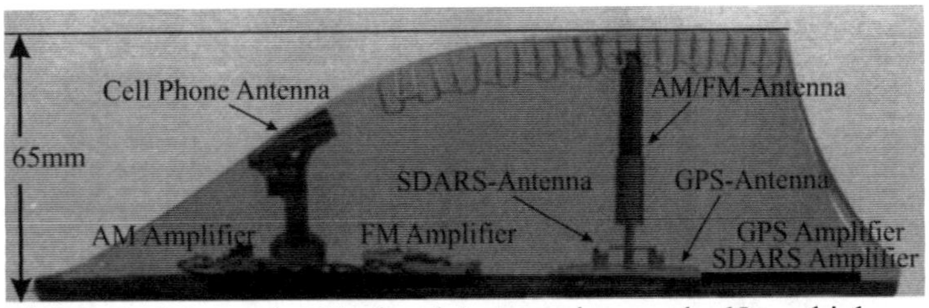

Figure 1. Automotive Hexband Antenna in an only 65mm high housing.

IOP Conf. Series: Materials Science and Engineering **44** (2013) 012001 doi:10.1088/1757-899X/44/1/012001

which has been extended in the form of a horizontally arranged meander to form a top-loaded antenna. This AM/FM antenna element is connected to separated amplifiers with capacitive high input impedance for AM and FM and is placed in the center of the SDARS/GPS antenna. As a result the performance of the SDARS/GPS antenna is not affected by the AM/FM antenna element at all. The SDARS and GPS antenna are realized in a combination of two table-formed ring structures with outer dimensions of only 39 mm x 39 mm x 10 mm [2] and are connected to separated low-noise amplifiers. The inner antenna for SDARS application has vertical elements for reducing the physical dimensions. A short cell phone antenna for AMPS and PCS1900 is integrated in the small housing, too.

3. AM/FM reception

The novel capacitive coupled helical antenna (CCHA) is only 56 mm high in total and consists of a copper helix with constant pitch, wound on a dielectric tube with a diameter of 6mm (figure 1). The lower end of the helix, isolated by a thin dielectric layer, is inserted in a cylindrical socket without having any galvanic contact with it. This socket is extended by a thin pin in order to be connected to the AM/FM amplifiers. The top end of the helix is continued to the right and left side with a bare wired meander structure, which is transparent for GPS and SDARS frequencies.

3.1. FM reception

The antenna impedance has resonant character in the FM range (87-108 MHz) and is displayed in figure 2. Therefore a specialized FM amplifier is necessary. The design principle of such a high input impedance amplifier is shown in figure 3. For FM, a two-stage amplifier is used, consisting of a GaAs HEMT transistor as source follower and a bipolar transistor employed in common base. In between both transistors a passive frequency flattening network is implemented in order to attenuate the resonances caused by the antenna in combination with the amplifier's capacitive high input impedance. An additional AGC network can be implemented here, too. The amplifier's gain is about 13dB, the noise figure less than 3dB. The output noise floor is below -5 dBμV at a bandwidth of 120 kHz. In order to keep off the flicker noise of the HEMT from the AM stage, a series resonant circuit with a low capacitive value C_s is applied in combination with a FM choke L. The dynamic range of the AGC comprises 27 dB and starts at 95dBμV input level. The k3 intermodulation product is 60dB below at each tone input level of 95 dBμV. The output VSWR is less than 2.

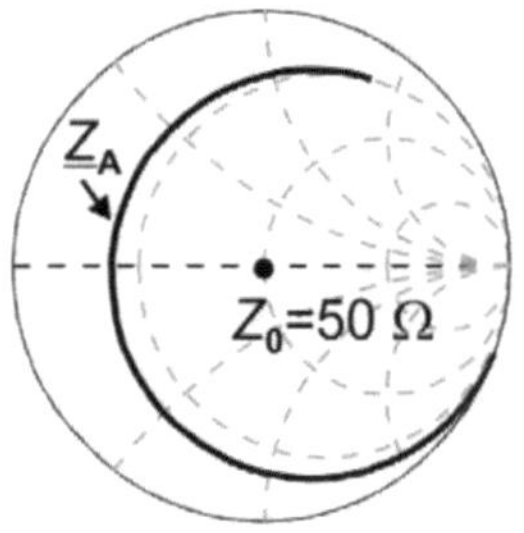

Figure 2. FM Antenna Impedance.

In order to evaluate the signal performance, field measurements are done with the active antenna mounted on top of a car above the rear windshield. The car is placed on a turntable and the values are averaged for a complete rotation of the car. In order to account for the particularities of the measurement setup, the results are compared to the ones obtained with a 900 mm passive monopole reference antenna mounted on a front fender of the car. The results plotted in figure 4 show the ratio of the signal delivered by the new antenna design to the signal delivered by the reference antenna for both polarizations. For horizontal polarization the mean value lies only -2.5 dB below the rod antenna and at -1.5 dB with vertical polarization from 87 to 108 MHz.

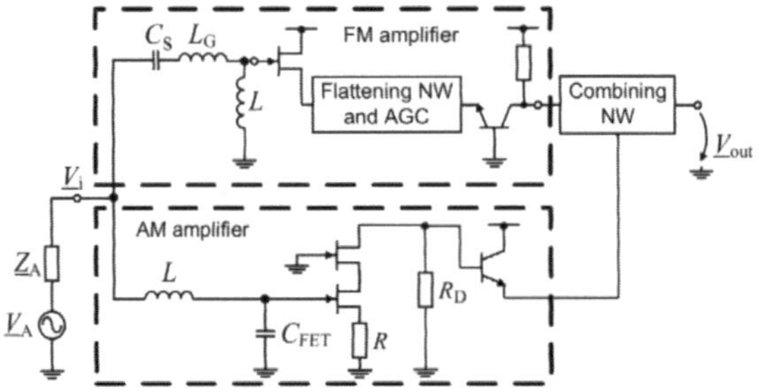

Figure 3. RF schematic of the AM/FM amplifier.

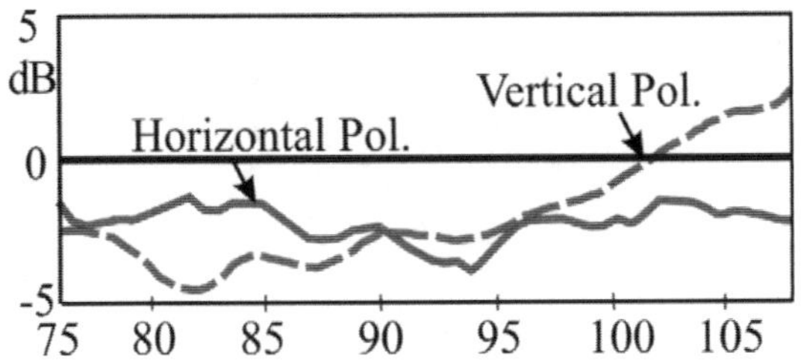

Figure 4. Field measurement in the FM range showing the ratio between the FM-Hexband antenna and the 900mm rod antenna for horizontal and vertical polarization.

3.2. AM reception

For a good AM performance it is obligatory to reduce capacitive loading by the amplifiers to an absolute minimum due to the low antenna capacitance. Therefore the AM amplifier is designed as a FET cascode followed by a buffer stage in order not to deteriorate the output signal level by using a long 50 Ohm coaxial cable to the receiver. At the input of the AM amplifier the FM choke L forms a low pass filter together with the input capacitance C_{FET} of the FET (figure 3). The filter prevents the full FM signal from reaching the AM stage. The designed AM amplifier fulfils the requirements [4] with respect to gain (R_D/R), noise floor, intermodulation behaviour and output compression point. Field measurements show that the signal levels achieved with the new antenna design is 2 to 3 dB in the LW and MW frequency range better than with a 900 mm long rod antenna.

4. SDARS and GPS reception

4.1. Antenna Principle

For the reception of Satellite Digital Audio Radio Services (SDARS) and Global Positioning Systems (GPS) a combination of two metal sheet antennas is used [3]. Both antennas have a ring structure at the top and vertical elements for reducing the physical dimensions [2]. This ring structures have an electro-magnetic length of exactly lambda at their main frequency. Therefore, the positive potential and the negative potential are always on opposite sides of the ring structures while the wave is circulating. In the geometric center of the ring antenna combination of this SDARS antenna and GPS antenna, the potential is always zero. This is the reason why it is possible, to integrate the AM/FM antenna structure in the geometric center of the SDARS-GPS antenna combination.

4.2. Design Principle

The SDARS- GPS antenna combination is realized in a combination of two table formed metal sheets with a ring at the top, vertical elements and capacitors to ground for reducing the physical dimensions. Therefore the volume with plastic socket is only 39 mm x 39 mm x 10 mm. The feed point for the SDARS antenna and the GPS antenna coupled with small matching networks and Low Noise Amplifiers (LNA) are positioned at the rear side of the circuit board. With perfect matching at 2339 MHz the inner SDARS antenna reaches a bandwidth of more than 50 MHz. The outer GPS antenna has also a bandwidth which exceeds the necessary bandwidth significantly. Cause of this metal sheet design of the SDARS-GPS combination in a plastic socket, the antenna combination is easy to fabricate, has a high reproducibility and low losses.

4.3. Antenna Measurement

The measurement results for the SDARS and GPS antenna were performed in an anechoic chamber with a distance of 5.45 m between the transmitting standard horn antenna and the Antenna Under Test (AUT). The AUT is mounted on a circular metallic plate with bent edges and a diameter of 1.2 m. In figure 5 the measurement results for the SDARS antenna are shown. The graph shows the gain for LHC polarization and RHC polarization in 36 azimuth cuts. The maximum gain for LHC polarization reaches 2.9 dBic in zenith. The SDARS antenna offers also a very high vertical gain. The linear average gain over phi, of the vertical polarization between theta = 75° and theta = 90° is – 0.51 dBil. The LNA offers a gain of 31 dB with a noise figure of 0.65 dB.

IOP Conf. Series: Materials Science and Engineering **44** (2013) 012001 doi:10.1088/1757-899X/44/1/012001

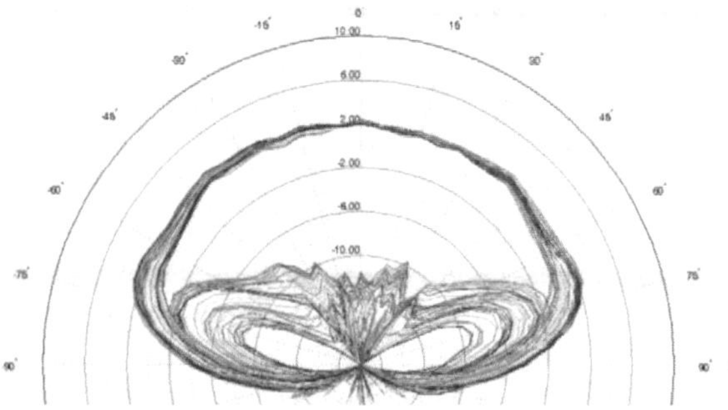

Figure 5. Gain of LHCP and RHCP at 2339 MHz of the SDARS antenna.

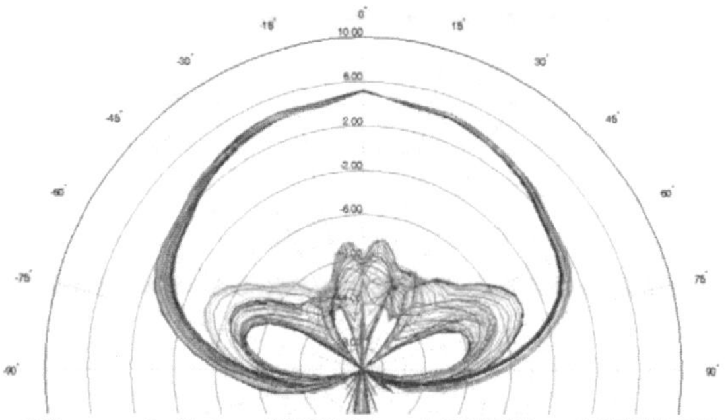

Figure 6. Gain of RHCP and LHCP at 1575 MHz of the GPS antenna.

Figure 6 shows the measurement results for the GPS antenna. The graph shows the gain for RHC polarization and LHC polarization in 36 azimuth cuts. The maximum gain for RHC polarization reaches 5.2 dBic in zenith. The LNA offers a gain of 30 dB with a noise figure of 1.5 dB. The efficiency of both antennas is about -1.5 dB.

5. Cell Phone Antenna

A short monopole antenna serves as cell phone antenna for AMPS and PCS1900 e.g. in the US. The lower part of this 2 band antenna is used for the higher PCS1900 frequency band, the upper part is connected via a printed series resonant circuit for the lower AMPS frequency band. In both frequency bands the VSWR is below 3 without applying a matching network.

6. Mutual Coupling

The mutual coupling between the antennas is for every used frequency range less than – 20 dB except the coupling between the AM/FM antenna and the cell phone antenna. A filter with very low capacitive values in front of the AM/FM amplifier suppresses the cell phone signals with 30 dB.

7. Summary

An automotive hexband antenna for AM/FM/GPS/SDARS and AMPS/PCS1900 cell phone in an only 65 mm high housing is investigated. Compared to a 900 mm long rod antenna the AM/FM antenna shows in field measurements for FM nearby the same performance and for AM even better signal levels. The SDARS-GPS antenna combination has been realized with the AM/FM structure in their center without influencing each other. The SDARS and GPS antennas achieve a high gain for satellite reception for both services. For cell phone, the VSWR reaches less than 3 in both frequency bands.

References

[1] A Negut, L Reiter, J Hopf and S Lindenmeier, "High-Impedance Amplifier for a Novel 14cm Short AM/FM Automotive Active Antenna", *Antennas and Propagation (EuCAP), 2010 Proceedings of the Fourth European Conference on*, April 2010, pp. 1 – 5

[2] J Kammerer and S Lindenmeier, "A compact car antenna with high efficiency for reception of heo- and geo- satellite signals" in *Antennas and Propagation (APSURSI), 2011 IEEE International Symposium*, July 2011, pp. 1205 – 1206

[3] J Kammerer and S Lindenmeier, "A new compact antenna combination with high efficiency for reception of SDARS- and GPS signals" in *Antennas and Propagation (APSURSI), 2012 IEEE International Symposium*, July 2012, pp. 1 – 2

[4] Verband der Automobilindustrie, VDA 240-200 specifications for car antennas, 01.01.2005

Design and Analysis of Inverted H Shape Dual Band Patch Antenna for Microwave Application

***M. Habib Ullah[1,2], M. T. Islam[1], J. S. Mandeep[1], N. Misran[1], M. B. I. Reaz[2]**
[1]Institute of Space Science (ANGKASA)
[2]Dept. of Electrical, Electronics and System Engineering
Faculty of Engineering and Built Environment
Universiti Kebangsaan Malaysia, Bangi, Selangor, Malaysia.

*E-mail: habib_ctg@yahoo.com

Abstract. Design and analysis of an inverted H shape printed dual band patch antenna has been presented in this paper. The proposed antenna has designed by 3-D full wave high frequency electromagnetic simulator HFSS and printed in low cost, durable epoxy polymer resin composite material substrate. The printed antenna prototype has been measured in a standard far field anechoic electromagnetic field measurement chamber. The measured results have analyzed by using computer aided plotting tool OriginPro 8.5. Measured -10 dB return loss bandwidth 3.25 GHz from 9.75 GHz to 13 GHz with peak gain 8.5 dBi have been achieved. The proposed antenna has obtained 0.63 dBi gain with 96% efficiency at lower band 10.3 GHz and 6.03 dBi gain with 84.2% efficiency at upper band 12.5 GHz. The almost steady radiation pattern makes the proposed antenna appropriate for part of X and Ku band applications. Moreover, the input impedance and current distribution along the radiating patch of the proposed antenna have also been analyzed in this paper.

1. Introduction

Owing to the ever increasing development of microwave technology there is a great need for compact, attractive, lightweight multi technology wireless device [1][2]. The demand for low profile compact antennas with multiband compatibility paved the way for the development of various types of patch antennas, monopole antennas, and planar inverted-F antennas (FIFAs) etc. The broadly used planar patch antennas have benefits like low profile, lightweight, simple structured and comparatively wider bandwidth than others. A numerous types of dual band antennas have studied, such as circle slot antenna [3], stacked patch [4], metamaterial branch line coupled [5], dual band dipole [6], slotted patch on ceramic material [7]etc. Most of these antennas are either relatively larger or does not have wide bandwidth.

A considerable amount of research effort has been given to develop compact, low cost, lightweight planar microstrip patch antennas for more than one application to integrate with multipurpose communication modules. It is obligatory for antenna researcher to develop compact low profile multiband planar antennas to meet the ever increasing demand of smart multi technology equipment

[1] Institute of Space Science, Universiti Kebangsaan Malaysia, 43600 Bangi, Selangor, Malaysia.
[2] Department of Electrical, Electronics and System Engineering, Faculty of Engineering and Built Environment Universiti Kebangsaan Malaysia, 43600 Bangi, Selangor, Malaysia.

[8]. A dual band dual-linearly-polarized proximity-coupled patch antenna was designed with dimension of 12.45 x 16 mm^2 [9]. A 10 x 16 mm^2 dual band-notched square monopole antenna was proposed for ultra wideband applications [10]. A diamond dual band antenna was designed for RFID application with overall dimension of 190 x 190 mm^2 [11]. However, the reported antennas are either larger in size or performance is not high in terms of bandwidth and/or gain.
In this paper, we propose a 5 mm long microstrip line fed inverted H shape slotted patch antenna for dual band operation with a compact size 12x15 mm^2 significantly smaller than the reported antennas. The measured results of the proposed antenna shows that -10 dB return loss bandwidth range from 9.75 GHz to 13 GHz (3.25 GHz) with peak gain 8.5 dBi and 0.63 dBi gain with 96% efficiency at lower band 10.3 GHz and 6.03 dBi gain with 84.2% efficiency at upper band 12.5 GHz.

2. Design and Configuration

The proposed antenna has been designed and analyzed by using finite element method (FEM) based 3D full wave high frequency electromagnetic simulator from ANSYS corp. [12,13]. Figure 1 shows the design layout of the slotted radiating patch of the proposed antenna. The optimized parameters of the proposed antenna are tabulated in table 1.

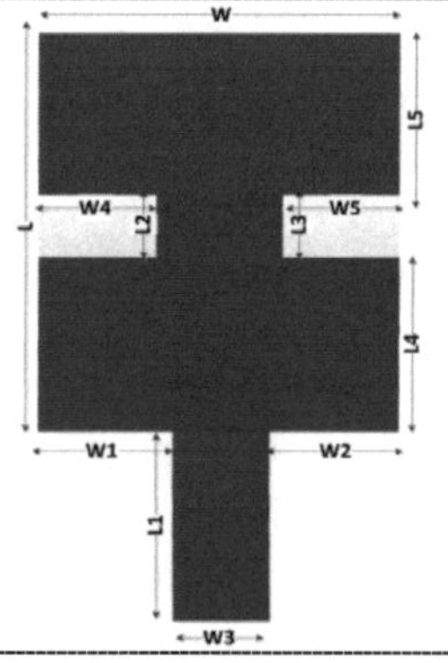

Figure 1. Design layout of the proposed antenna.

Figure 2. Photograph of the proposed antenna prototype.

Table 1. Design specifications of the proposed antenna.

Parameter	Value (mm)	Parameter	Value (mm)
L	15	W	12
L1	5	W1	4.5
L2	2	W2	4.5
L3	2	W3	3
L4	6.5	W4	4
L5	6.5	W5	4
L	15	W	12

The proposed inverted H shape radiating patch has obtained by cutting slots from conventional rectangular shape. By cutting slots the radiating patch lakes longer path around the slots for the flowing current till reach the opposite edge. The desired resonant frequency and the bandwidth has achieved by introducing slots on the radiating patch. Figure 2 shows the photograph of the proposed antenna prototype.

3. Results and Analysis

The performance of the proposed antenna has been analyzed and optimized by using finite element method (FEM) based high frequency three dimensional full wave electromagnetic simulator Ansoft's HFSS and plotted by using OriginPro 8.5. The results of the proposed antenna prototype have been measured in a standard far field anechoic measurement chamber.

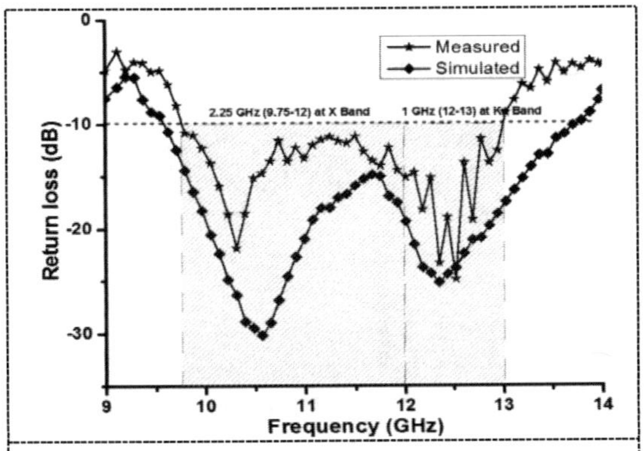

Figure 3. Simulated and measured return loss of the proposed antenna prototype.

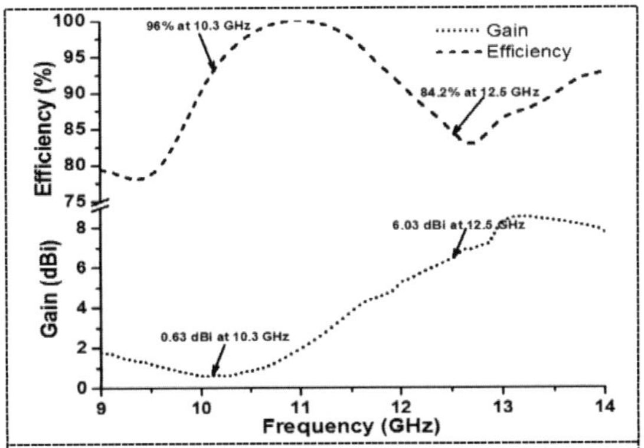

Figure 4. Gain and radiation efficiency of the proposed antenna prototype.

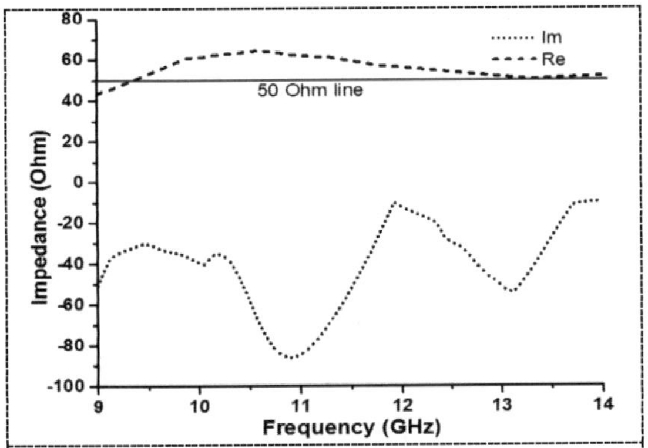

Figure 5. Input impedance of the proposed antenna.

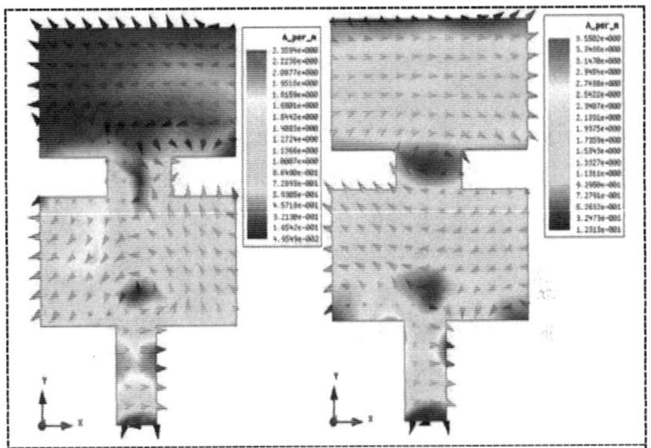

Figure 6. Current distribution along the radiating patch of the proposed antenna.

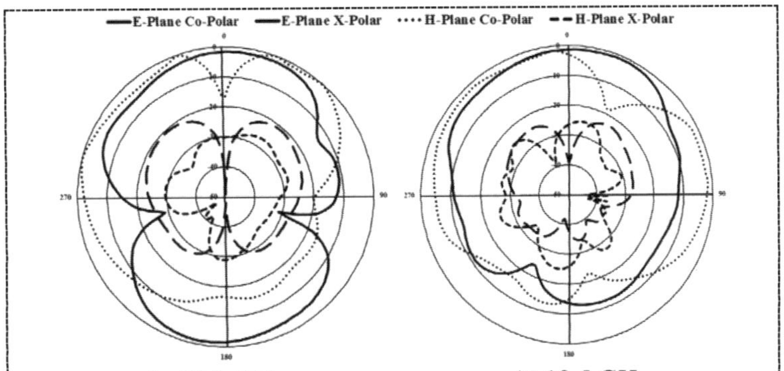

At 10.3 GHz At 12.5 GHz

Figure 7. Radiation pattern of the proposed antenna prototype.

The simulated and measured return loss of the proposed antenna has shown in figure 3. It can be clearly observed from the measurement that, the 10 dB bandwidth of 3.25 GHz from 9.75 GHz to 13 GHz which shows that the lower resonance shifted from 10.5 GHz to 10.3 GHz and the upper resonance shifted from 12.3 GHz to 12.5 GHz. Figure 4 shows the gain and efficiency of the proposed antenna. It is evidently noted that, 0.6 dBi and 6.03 dBi gain has been achieved with the radiation efficiency 96% and 84.2% at the lower and the upper resonances respectively. The input impedance of the proposed antenna is presented in figure 5. The real part of the impedance was optimized to be as close as possible to a 50 ohm line. Current distribution along the radiating patch of the proposed antenna has shown in figure 6. It can be clearly seen that, the intensity of current distribution at upper frequency is higher than lower resonant frequency. The E-H plane radiation pattern of the proposed antenna has shown in figure 7. It has shown that, 3dB beam width at 10.3 GHz the 56° (280°-34°) and 38° (286°-4°) at 12.3 GHz. The almost steady radiation pattern at both resonant frequencies makes the proposed antenna suitable for X and Ku band satellite applications.

4. Conclusion

A compact, low profile microstrip line fed inverted H shape 15 mm x 12 mm slotted patch antenna has been design and prototyped on low cost durable FR4 substrate operating frequency ranges from 9.75 GHz to 13 GHz. The result shows that, 3.25 GHz impedance bandwidth (return loss <-10dB) with 8.5 dBi peak gain is achieved. 96% and 84.2% radiation frequency are testified at both resonant frequencies 10.3 GHz and 12.5 GHz. The symmetric and almost steady E-H plane radiation patterns are observed. Furthermore, input impedance and current distribution along the radiating patch are analyzed.

[1] Azim R, Islam M T and Misran N 2011 *IEEE Antennas Wirel. Propag. Lett.* **10** 1190
[2] Mobashsher A T, Islam M T and Misran N 2010 *IEEE Antennas Wirel. Propag. Lett.* **9** 653
[3] Emadian R, Mirmozafari M, Ghobadi C and Nourinia J 2012 *Electron. Lett.*, **48** 356
[4] Ullah M H, Islam M T, M. S. Jit, and N. Misran 2012 *J. Intell. Mater. Syst. Struct.* **23** 1827
[5] Jung Y K and Lee B 2012 *IEEE Trans. Antennas Propag.* **60** 786
[6] An W X, Wong H, Lau K L, Li S F and Xue Q 2012 *IEEE Trans. Antennas Propag.* **60** 1592
[7] Ullah M H, Islam M T and Mandeep J S 2012. *Microwaves & RF,* **51** 50
[8] Tiang J J, Islam M T, Misran N and Mandeep J S 2011 *Prog. Electromagn. Res.* **120**, 499
[9] Veysi M, Kamyab M and Jafargholi A 2011 *IEEE Antennas Propag. Mag.* 53 90 –96, 2011.
[10] Mehranpour M, Nourinia J, Ghobadi C and Ojaroudi M 2012 *IEEE Antennas Wirel. Propag. Lett.* **11** 172
[11] Sabran M I, Rahim S K A, Rahman A Y A, Rahman T A and Nor M Z M 2011 *IEEE Antennas Wirel. Propag. Lett.* **10** 979
[12] Ullah M H, Islam M T, Mandeep J S and Misran N 2013 *Appl. Phys. A-Mater. Sci. Process.* **110** 199
[13] Ullah M H, Islam M T, Mandeep J S, Misran N and Nikabdullah N 2012 *Electronics and Electrical Engineering* **123** 75

IOP Conf. Series: Materials Science and Engineering **44** (2013) 012003 doi:10.1088/1757-899X/44/1/012003

Design and Development of a Small Compact Ultra Wideband Antenna

Qurratulain[1], Neela Chattoraj[2]
[1]Ph.D Scholar, Birla Institute of Technology, Mesra, Ranchi, India.
[2]Professor, Birla Institute of Technology, Mesra, Ranchi, India.

E-mail: annie13ahmad@gmail.com

Abstract. In this paper the design of a novel and compact U slot UWB printed monopole microstrip antenna of size 19.2 x 28.8 mm^2 has been presented for wireless applications. The planar, small and thin UWB antenna design consists of a U slot radiator fed by a single 50Ω microstrip line with truncated ground plane is excited by a coaxial SMA connector. The simulations are done using the Ansoft High Frequency Structure Simulator (HFSS) software tool. The simulated results of impedance bandwidth are well supported by measurement. The measured group delay and radiation pattern results are also presented and performance of the antenna is analyzed/discussed.

1. Introduction

In the last decade, Ultra-wideband (UWB) has come up as a revolutionary and contemporary wireless technology which has generated a great deal of interest for use in the industry and academia. The rapid progress of UWB as a high data rate wireless communication technology has mainly been spurred on by the release of a bandwidth of 7.5 GHz (3.1 GHz - 10.66 GHz) for UWB applications by the Federal Communications Commission (FCC) [1,2].

As is the case in any conventional wireless communication systems, an antenna also plays a very fundamental role in UWB systems. On the other hand the challenges faced in designing a UWB antenna are many more. A good candidate for UWB applications are printed monopole antennas due to their compactness, light weight and simple structure [1, 3, 4, 5, 6, 7].

The design of a tapered U slot printed monopole microstrip antenna is reported in ultra wideband for wireless applications in this paper. First the planar PCB antenna design is introduced. Also the various design considerations and dimensions are summarized. Finally the simulated and measured results have been compared and in the process a novel antenna is realized and fabricated.

2. UWB Antenna Design

The planar, very small, thin UWB antenna is etched on a 19.2mm x 28.8mm FR4 substrate having a relative permittivity of 4.4 and a substrate height of 1.6mm. The geometry of the proposed antenna is shown in figure 1 and the photograph of the fabricated antenna is shown in figure 2. The U slot radiating patch and transmission line are connected by tapered edges to obtain better impedance matching. A partial ground plane having the length of 9.2mm is used on the other side of the substrate.

It consists of a slot of dimensions 2.8mm x 1.2mm just behind the transmission line to improve the bandwidth.

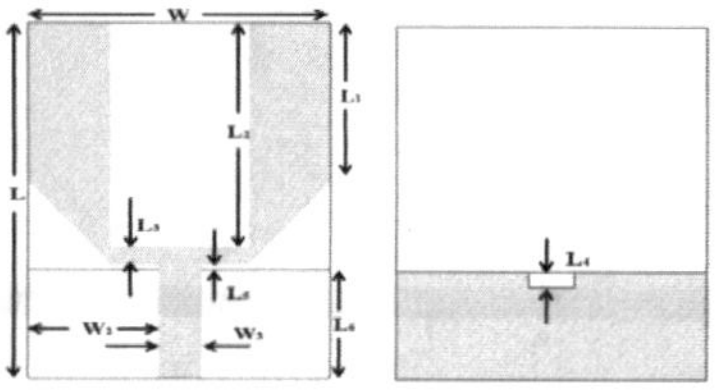

Figure 1. Geometry for the proposed antenna.

Figure 2. Fabricated UWB Antenna.

The resonant length of a microstrip patch can be obtained by using simple relations of the effective relative dielectric constant as a function of the substrate parameters and the operating frequency as follows [8]

$$\epsilon_{\text{reff}} = \frac{\epsilon_{\text{reff}}+1}{2} + \frac{\epsilon_{\text{reff}}-1}{2}\left[1+12\frac{h}{W}\right]^{\frac{-1}{2}} \quad (1)$$

$$W = \frac{c}{2f}\sqrt{\frac{2}{\epsilon_r+1}} \quad (2)$$

$$L_{\text{eff}} = \frac{c}{2f\sqrt{\epsilon_{\text{reff}}}} \quad (3)$$

$$\Delta L = 0.5h \quad (4)$$

$$L = L_{eff} - 2\Delta L \quad (5)$$

where c is the speed of light in free space, L and W are the length and the width of the resonant patch antenna respectively. By simulating different patches with substrate parameters, it is possible to obtain the appropriate values as the patch does not have a ground plane on the other side of the substrate as shown in figure 1.The center frequency used in the simulations is 6 GHz. For this operating frequency and substrate parameters which are discussed before, the dimensions of the patch are calculated to be W = 15.21mm and L = 11.98mm. These dimensions represent the starting point for the present design of the UWB antenna as shown in figure 1. The parameters of the proposed antenna are given in Table I.

Table 1. Detailed parameters for the proposed UWB antenna.

Serial No.	Symbols	Size(mm)
1	L	28.8
2	W	12.8
3	L_1	18.16
4	L_2	1.04
5	L_3	1.2
6	L_4	0.4
7	L_5	9.2
8	L_6	19.2
9	W_1	5.2
10	W_2	8.2
11	W_3	2.8

Some modifications are introduced on this patch to improve its operating bandwidth. The first one is to taper the patch near the feeding microstrip line. Due to the tapering effect of that part connected to the microstrip line, the dimensions of the patch are modified through simulations to be matched with the line such that the length L of the patch is changed to be 12.8mm and the width W is changed to be 19.2mm as mentioned in table 1. Using these dimensions with tapering length L_{taper} = 6.4mm are adequate to introduce an UWB patch antenna that operates in the frequency range from 4.1 to 14.0 GHz as is shown by the simulated result in the following section.

3. Results and Discussion

Figure 3 shows the variations of the measured and simulated results of impedance bandwidth (VSWR< 2) for the proposed UWB antenna. The measurements were done using vector network analyzer (VNA, PNA N5230A, Agilent Technologies).The results show that the proposed fabricated antenna achieves an impedance bandwidth (VSWR< 2) from 3.4 – 14 GHz . Some differences in the simulated and measured return loss curves (S_{11} plots) are seen. The main reason is that measurement is done in a non-controlled environment.

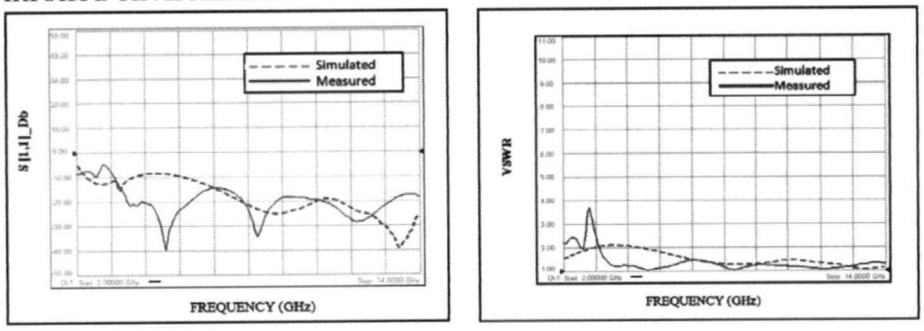

Figure 3. Simulated and measured return loss and VSWR curves.

Figure 4 show the simulated radiation patterns of the proposed antenna at 4.5, 6.5, 8.5 and 10.5GHz.The far field radiation patterns measured using C-band (4-8GHz) and X-band (8-12.4GHz) Microwave benches at 6.5, 8.5 and 10.5GHz are plotted in Figure 5.

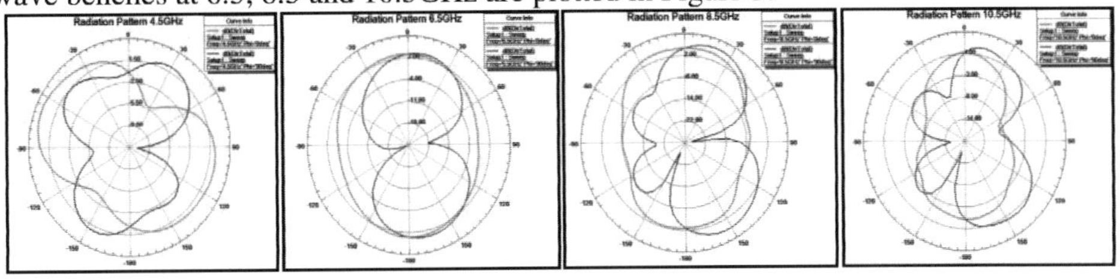

Figure 4. Simulated Radiation patterns at 4.5, 6.5, 8.5 &10.5GHz.

It is clearly seen that the radiation patterns of E-plane are monopole like and H-plane radiation patterns show almost omni-directional characteristics, especially at lower frequencies. However the radiation patterns start to change in higher frequencies and UWB antenna becomes directive. At high frequencies (i.e. from 7GHz onwards) because of spurious radiation side lobes appear as seen in figure 4.

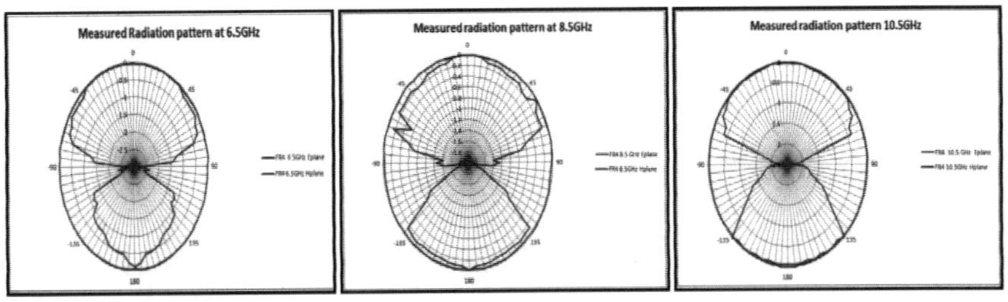

Figure 5. Measured Radiation patterns at 6.5,8.5 & 10.5GHz.

The simulated antenna gain varies from 4.4dB to 6dB over the operating UWB frequency range as shown in figure 6. Another important parameter in UWB antenna design is the group delay. From the measured results the antenna group delay is approximately constant and within the frequency band of interest as shown in figure 7.

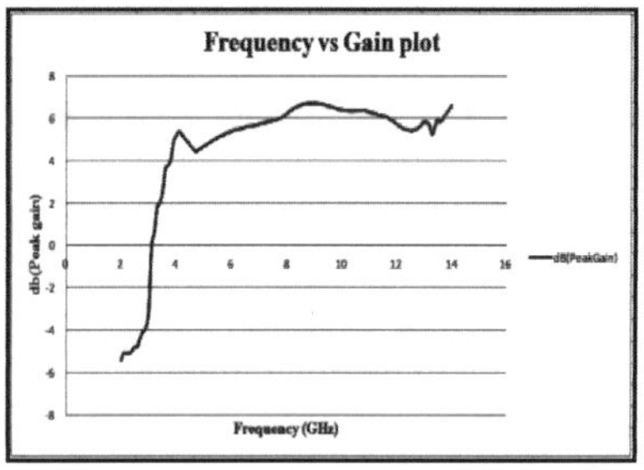

Figure 6. Simulated antenna gain.

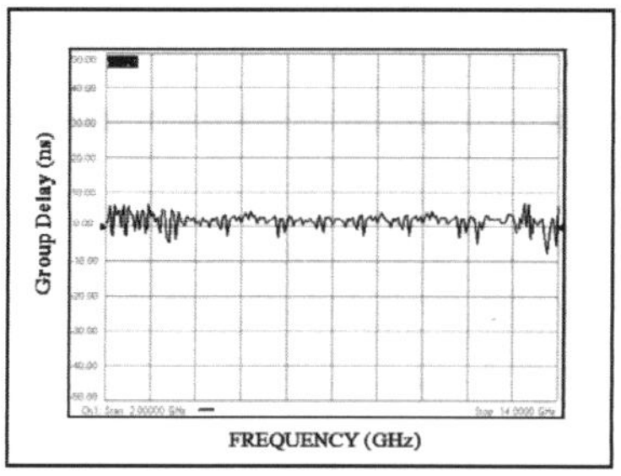

Figure 7. Measured group delay for fabricated UWB antenna.

4. Conclusion

A new small UWB antenna has been designed, simulated, measured and fabricated. The simulation results obtained by Ansoft HFSS software show good agreement with the measured results. Group delay which is more meaningful parameter to show good time domain characteristic in UWB antenna is investigated and the variations obtained are also good and within the acceptable limit. To miniaturize UWB antenna, tapering and truncated ground plains are used. It is seen from the measured result that very large bandwidth is obtained for the proposed antenna which can be used in UWB communication systems.

5. Acknowledgement

The authors would like to acknowledge and thank the support of Birla Institute of Technology (Mesra) for providing the facility and equipment for this work.

References

[1] Schantz H 2005 The Art and Science of Ultra Wideband Antennas (Norwood MA: Artech House Inc)

[2] Federal Communications Commission Revision of Part 15 of the Commission's Rules Regarding Ultra-Wideband Transmission System from 3.1 to 10.6 GHz 2002 *Federal Communications Commission* FCC (Washington DC: ET-Docket) pp 98–153

[3] Lim E G, Wang Z, Lei Chi-Un, Wang Y and Man K L 2010 Ultra Wideband Antennas – Past and Present *IAENG International Journal of Computer Sciences* **37** no. 3

[4] Khalilpour R, Nourinia J and Ghobadi C 2010 An Optimized Monopole Microstrip Antenna with Gradual Steps for Ultra Wideband Applications *Progress In Electromagnetic Research Symposium Proceedings* pp 1072-1076

[5] Song H W, Park J K, and Yoo J H 2008 A Novel Ultra-Wideband Monopole Antenna With Two Symmetrical Strips *Microwave and Optical Technology Letters* **50** no. 11 pp 2845 - 2848

[6] Yin X C, Ruan C L, Ding C Y and Chu J H 2008 A Planar U Type Monopole Antenna For UWB Applications *Progress In Electromagnetic Research Letters* **2** pp 1–10

[7] Lee Y, Hong S, Kim J and Choi J 2010 Design of an Antenna with Near Omni-directional H-Plane Radiation Pattern over Ultra-wide Bandwidth *ETRI Journal* **32** no. 1

[8] C. Balanis 1982 Antenna Theory; Analysis and Design (second edition Wiley and Sons, New York)

IOP Conf. Series: Materials Science and Engineering **44** (2013) 012004 doi:10.1088/1757-899X/44/1/012004

Circular Antenna Array Pattern analysis using Radial Basis Function Neural Network

B Rama Sanjeeva Reddy[1], D Vakula[2] and N V S N Sarma[3]

[1] Research Scholar, National Institute of Technology, Warangal, India
[2] Assistant Professor, National Insititute of Technology, Warangal, India
[3] Professor, National Insititute of Technology, Warangal, India

E-mail: sanjeev.reddy@nitw.ac.in

Abstract. A method is proposed to design circular antenna array for the given gain and beam width using Artificial Neural Networks. In optimizing circular arrays, the parameters to be controlled are excitation of the elements, their separation, lengths and the circle radius. This paper deals about finding the parameters of radiation pattern of given uniform circular antenna array. Initially, the network is trained with a set of input-output data pairs. The trained network is used for testing. The training data set is generated from MATLAB simulation with number of elements N=5, 10, 15 and 20 elements of uniform circular array, respectively, distributed over a given circle, assuming 20 training cases. The number of input nodes, hidden nodes and output nodes are 20, 20 and 1, respectively. Predicted values of the neural network are compared with those of MATLAB simulation results and are found to be in agreement. This work establishes the application of Radial Basis Function Neural Network (RBFNN) for circular array pattern optimization. RBFNN is able to predict the output values with 97% of accuracy. This work proves that RBFNN can be used for circular antenna array design.

1. Introduction

Circular array antennas are very popular due to several advantages such as an all-azimuth scan capability (i.e., the array can perform a 360^0 scan around its center), and ability to keep the beam pattern invariant. Circular array antennas contain many circular rings of different radii and with different number of elements. These arrays have several advantages, including flexibility in array pattern design and synthesis, in both [1] narrowband and broadband beam-forming applications. Circular arrays [2] are also favored in direction-of-arrival applications, since they provide almost invariant azimuth-angle coverage. This work establishes the application of RBFNN for circular array design optimization in antenna arrays. Its applications [6] span radio direction finding, air and space navigation, underground propagation, radar and sonar system.
The array factor for circular array is given in equation (1)

$$AF(\theta,\phi)=\sum_{n=1}^{N} I_n e^{jka[\sin\theta\cos(\phi-\phi_n)-\sin\theta_0\cos(\phi_0-\phi_n)]} \qquad (1)$$

where θ_0 and ϕ_0 are the angles of the main beam, a is the radius of the circular array and I_n are the excitation coefficients of the elements.

In this work, a circular antenna array is considered, attention being on gain and beam width parameters. Artificial Neural Networks are implemented in designing a circular antenna array The training data set is generated from MATLAB simulations with number of elements N=5, 10, 15 and 20 elements distributed along a given circle, assuming 20 training cases; the measured error is calculated from comparison of simulated values with the values generated using a Radial basis functions neural network.

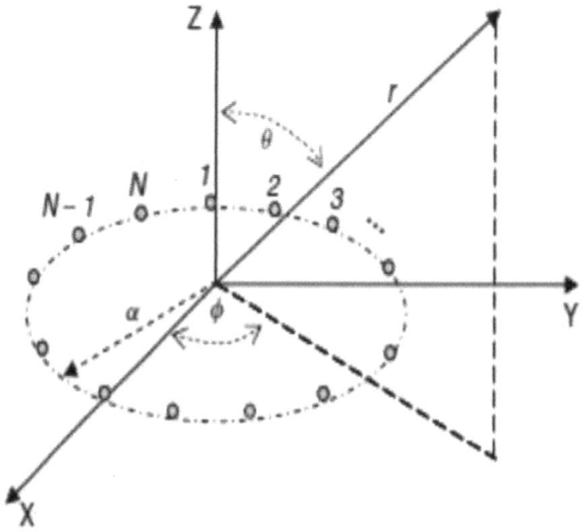

Fig1: Geometry of N-element circular array.

Figure 1 represents the geometry of N-element circular array antenna with the change in azimuth and elevation angles.

2. Artificial Neural Networks

Artificial Neural network is an application that alters certain variables in response to a set of corresponding input and output patterns. Beginning with an initial set [5] of internal values, the network modifies these quantities in order to find a position of "best fit," thereby generating from the input patterns their expected results. The ability [7] of these networks to generalize relationships between inputs and outputs is a key to their effectiveness. RBFNN typically have three layers of neurons, namely input, hidden and output, which are fully interconnected as shown in Figure 2.

The first layer is composed of input nodes. An RBF network is a feed forward neural network with one hidden layer, with an RBF node function at each hidden node; the weight vector from the input layer to a hidden node is identified to the location of the center of the RBF for that node. The number of nodes, n, is equal to the dimension of input vector.

The second layer is a hidden layer composed of multivariate Gaussian nonlinear units that are connected directly to all of the input layer nodes. Each neuron in the hidden layer operates at the Gaussian transfer function. RBF network with a sufficiently large number of nodes can approximate any real multivariate continuous function on a compact set.

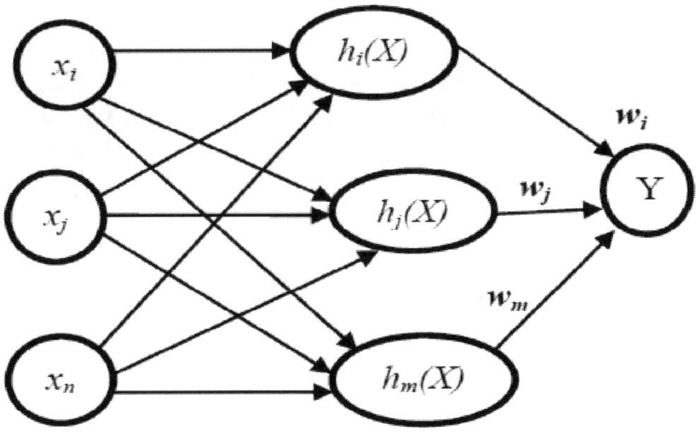

Fig2: RBF Network Structure

3. Algorithm steps

- Input parameters of the uniform circular array are number of elements, element excitation, element spacing, and the element radius.
- Determine the gain and beam width output parameters of array with ON/OFF mode of number of switching elements using MATLAB simulation
- Training set consists of number of ON elements, gain and beam width as the inputs.
- Excitation for corresponding element is determined for training the neural network
- 20 data sets are considered for the network and out of which 15 sets of data samples are taken for training and 5 random data samples are considered as testing sets.
- Compute the predicted values with the actual data values
- Compute the change in mean average percentage error (measured error) for the technique

4. Results and Discussion

Consider a circular array with 5, 10, 15 and 20 elements respectively, the output parameters like in [4], gain and half power beam widths, being calculated. Figure 3 shows the radiation pattern for 5-element circular array generated using MATLAB simulation.

Figure 4 gives the resultant power pattern along φ cut corresponding to the circular path which finds the half power beam width (HPBW) for a 5-element circular array.

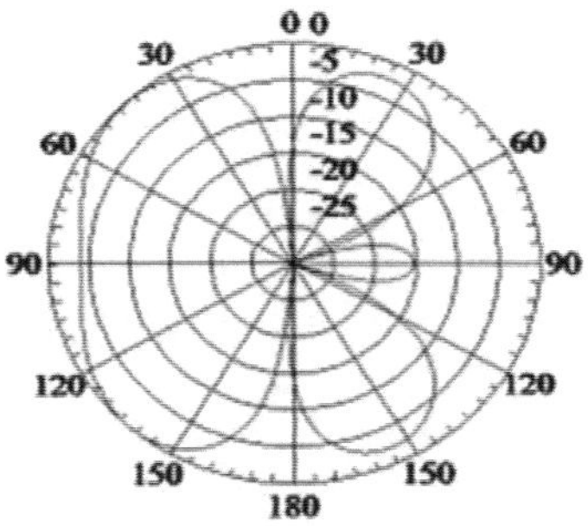

Fig3: Principle plain amplitude pattern of circular array for N=5

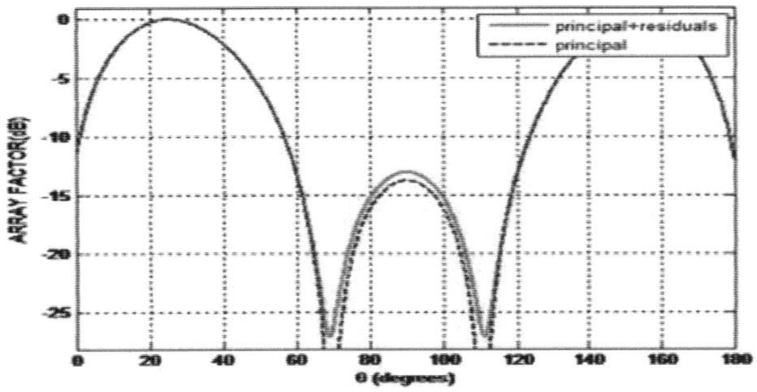

Fig4: Power pattern corresponding to circular array

To establish the validity of the procedure, a circular array with 5 isotropic elements with uniform excitation and uniform spacing is considered. Table I gives the list of simulated output parameters of this circular array using MATLAB.

Table I. Parameters and Measured Error of Circular array (θ=0, φ=90)

Input Parameters		**Output Parameters (Simulated)**		**Predicted number of elements using RBFNN**	**% Measured error**
Elements	***Radius***	***Gain***	***HPBW***		
5	0.5	6.0677	42.0203	5.0621	1.242
10	0.5	6.145	0.5	11.249	1.249
15	0.5	6.145	33.255	15.18	1.241
20	0.5	7.5258	33.255	20.25	1.244

Mean average percentage error (MAPE) is calculated as (Observed value-Predicted value) / Predicted value, and the measured error calculated for the trained network data is as shown in Table I. Given simulated data is taken as the inputs for the training network and the number of elements are predicted using RBFNN algorithm. The samples are given as inputs to the neural network. Each network [5] is trained with 10 test patterns. During the training it is observed that only 20 input nodes are sufficient. Using more input nodes leads to more accurate representation of the pattern but will also require a much larger network and hence the computational time will increase. With the present architecture, the results indicate that for 97% of all cases presented, the network showed the correct result.

5. Conclusion

Patterns of circular antenna arrays are discussed with the selection of the elements using a Neural Network. Circular array pattern design and analysis is essential for multiple beam forming technique for antenna design. ANNs have better learning ability, generalization, parallel processing and error endurance attributes that lead to perfect solutions in applications where one needs to model the nonlinear mapping of complex data. This approach makes use of Neural Network that can be trained for any number of elements, spacing and excitation. Once the network is trained, it can find the parameters with respect to the input. Gain and half power beam width are the output parameters. Number of elements and radius of the ring are the input parameters for the training network. RBFNN is able to predict the output values with 97% of accuracy. The designed model of RBFNN has given quick speed of convergence and improved accuracy. This work proves that RBFNN can be used for circular array design.

References

[1] R W P King Super gain antennas and the Yagi and circular array 1989 *IEEE Trans. Antennas and Propogat.* vol **37** No 2 178-186
[2] M Khodier and M Al-Aqeel 2009 Linear and circular array optimization : A study using particle swarm intelligence *PIER B* vol **15** 347-373
[3] D Vakula and N V S N Sarma 2006 A generalized method to diagnose faults in antenna arrays using neural networks *AMTA* 290-292 USA
[4] C A Balanis 1997 *Antenna Theory and Design* John Wiley & Sons Newyork 321-330
[5] S Haykin 1999 *Neural Networks: A Comprehensive Foundation* 2nd edition Prentice Hall
[6] R Fallahi and M. Roshandel 2007 Effect of mutual coupling and configuration of concentric array antenna on the signal to interference performance in CDMA Systems *PIER* **76** 427-447
[7] Q J Zhang and K C Gupta 2000 *Neural Networks for RF and Microwave Design* (Boston: Artech house)

IOP Conf. Series: Materials Science and Engineering **44** (2013) 012005 doi:10.1088/1757-899X/44/1/012005

Koch fractal boundary patch over reactive impedance

Venkateshwar Reddy V[1] and N.V.S.N Sarma[2]

Electronics and Communication engineering Department, NITW, INDIA

E-mail: [1] vvreddy2005@gmail.com, [2] sarma@nitw.ac.in

Abstract. This paper describes the enhancement of bandwidth and miniaturization for patch antennas. Introduction of fractal structure (Square Koch) over reactive impedance surface (RIS) is used to enhance impedance bandwidth while minimizing the patch size. Comparison has been made with those of a single-layer (sub1) antenna and the corresponding dual-layer (RIS) antenna. Approximately double the impedance bandwidth is achieved with the proposed RIS Square Koch antenna 1 when compared with Square Koch antenna 1without RIS. There is a 55 % reduction in the patch size. The simulated results indicate that the presented antennas provide gain of about 2.5dBi over the entire band of frequencies.

1. Introduction

Compact and multiband antennas are hot favourites in personal and mobile communications. Fractals are used in the design of microstrip antennas for multiband operation and miniaturization [1]. Metamaterial-based structures have been extensively studied for antenna applications in the past decade and in particular the high-impedance surfaces (HIS) and the reactive impedance surfaces (RIS). When associated with radiating elements, RIS can be used to enhance their bandwidth as well as to reduce their size (for a given resonant frequency), in comparison to radiating elements printed on conventional grounded dielectric substrates. Based on the work of Sarabandi *et al.* [2, 3], a square microstrip patch antenna printed on an RIS substrate has been designed at ISL [4]. Bandwidth enhancement and size reduction have been demonstrated numerically and experimentally, in comparison with the same patch printed on the same substrates but without the RIS lattice.

Antenna miniaturization utilizing a patch antenna printed on a metal backed high dielectric substrate has been attempted previously [5]. The drawbacks of this approach are twofold : 1) there is a strong electromagnetic coupling between the patch and the PEC surface backed by the high permittivity medium, which traps a significant portion of EM energy in the near fields resulting in a low antenna efficiency, narrowband characteristics, and in some cases an unwanted radiation pattern caused by surface and leaky waves, 2) the characteristic impedance in a high permittivity medium is rather low which creates difficulty in impedance matching of the antenna. These challenges can be successfully overcome by the novel design of patch antenna over a Reactive Impedance Surface (RIS) [6]. The RIS has the minimum interaction with the patch and dramatically enhances antenna performance metrics, namely, size, bandwidth, and efficiency.

Previous works on the RIS are limited to only conventional patches like square, square with slots etc. Here in this communication, the possibility to design bandwidth-enlarged linear polarized fractal antenna on RIS is reported.

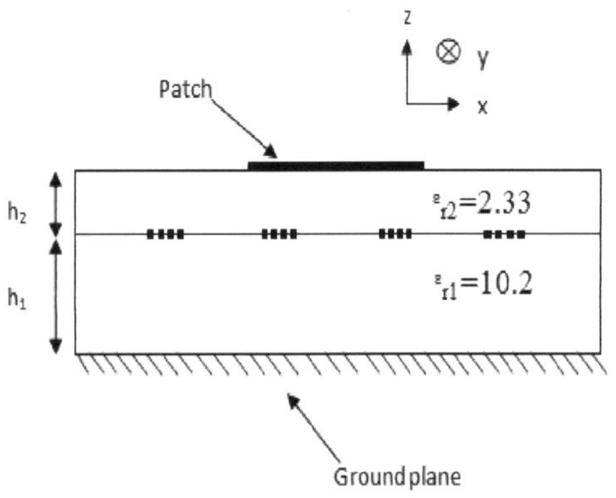

Figure 1.a. Side view of RIS based Square Koch fractal antenna.

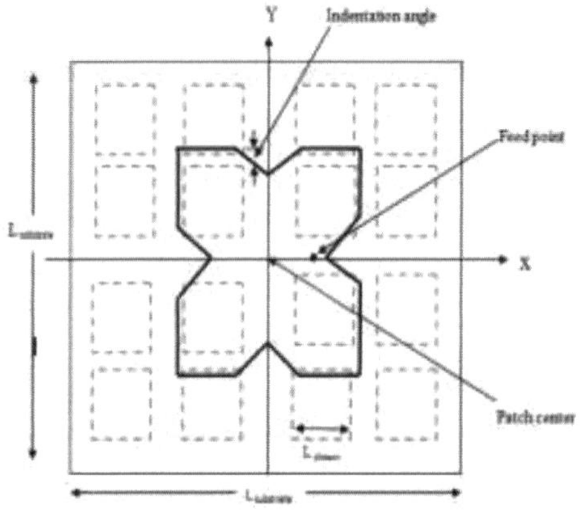

Figure 1.b. Top view of RIS based Square Koch fractal antenna.

2. Antenna Geometries

The RIS-based structures considered in this work are represented in Fig. 1. They consist of a radiating element printed on the top of a dual-layer substrate in which the ground plane lies at the back side of the structure and the lattice of square elements (RIS) is printed at the interface between both dielectric layers. These RIS elements are periodically spaced along *x*- and *y*-axes, thus leading to a symmetrical design (the RIS arrangement is centered with respect to the patch center). The substrate size equals $L_{substrate}$=48 mm, the thickness and permittivity of the bottom and top substrates are respectively h_1=6 mm, h2=3.2 mm, ε_{r1}=10.2 and ε_{r2}=2.33. The dimensions of the RIS elements are $L_{element}$= 6 mm and the spacing between the elements is 4 mm.

The geometry of the square Koch fractal patch is given in Fig 1.b. Here for single layer, end to end length of the fractal curve is 36 mm and for dual layer fractal curve, the length of each side is 24 mm. The antenna is fed along the X-axis at a distance from the origin where the antenna impedance is about 50 ohms. The indentation angle varies from 10^0 to 40^0 with a step size of 10^0. The finite-element-method (FEM) based software Ansoft HFSS is used to carry out all the simulations.

3. Results and Discussions

It can be seen from the figs 2 and 3 that as the iteration increases, the resonance frequency is shifted left towards the origin. It is because of increase of electrical length of the patch, as given in the literature the resonance frequency is inversely proportional to electrical length of the patch. Here sub1 indicates single layer with relative dielectric constant 2.33. Radiation pattern and current distribution of designed antennas are depicted in fig 4. From fig 4.a it gives that gain is 2.5dBi and from fig 4.b it shows current distribution more at Koch fractal curve edges.

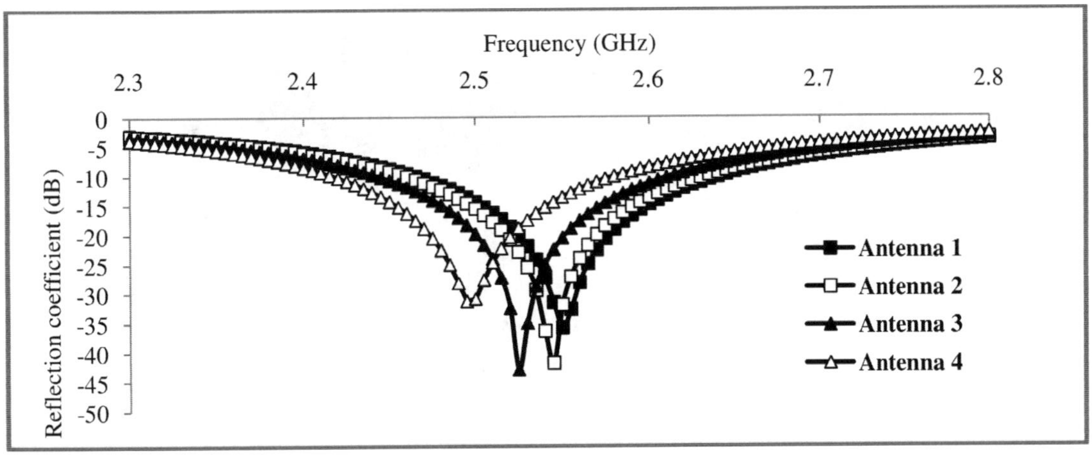

Figure 2. S_{11} **Characteristics of square Koch fractal boundary sub 1(without RIS) patch antenna.**

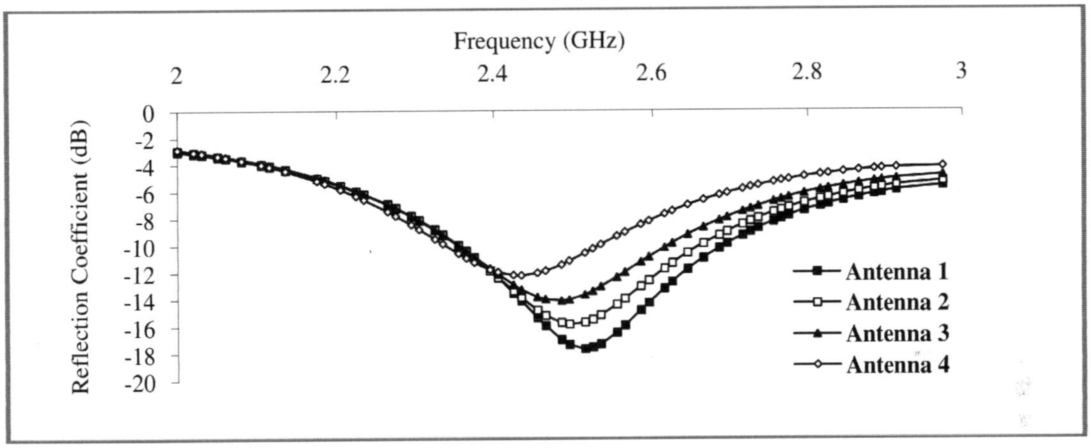

Figure 3. S_{11} **Characteristics of square Koch fractal boundary RIS patch antenna.**

Table 1. Comparison of sub1 and RIS return loss characteristics for different iteration angles.

Antenna	Indentation angle	Layer	Patch size(mm)	Min (S_{11}) (dB)	S_{11} BW(10dB) (MHz)
Antenna 1	10^0	Sub1	36	-35.865	175
		RIS	24	-17.7	330
Antenna 2	20^0	Sub1	36	-41.82	170
		RIS	24	-16	310
Antenna 3	30^0	Sub1	36	-42.94	165
		RIS	24	-14.2	265
Antenna 4	40^0	Sub 1	36	-31.28	160
		RIS	24	-12.5	200

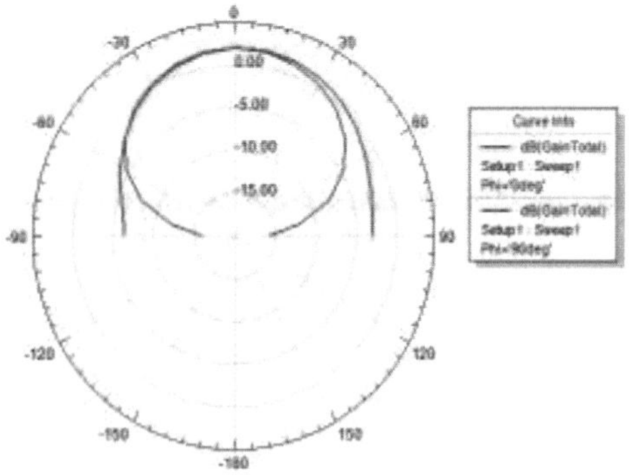

Figure 4.a. Antenna 4 sub1 radiation pattern

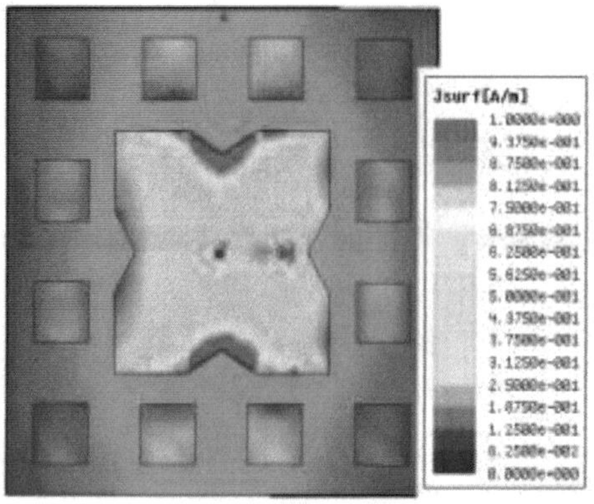

Figure 4.b. Current distribution of RIS antenna 3 at 2.485 GHz

4. Conclusion

Investigations on the performance and characteristics of fractal antennas on Reactive Impedance Substrate (RIS) have been presented. The RIS structures are compared with "conventional" structures (i.e. without RIS lattice). The overall impedance bandwidth of RIS antenna 1 is approximately double when compared with sub 1 antenna 1. Antenna miniaturization is possible with RIS antennas.

References

[1] Douglas H Werner and Suman Ganguly 2003 An overview of Fractal Antenna Engineering Research *IEEE Antenna and Propagation Magazine* vol 45 no1 p 1101-1113

[2] H. Mosallaei and K. Sarabandi 2004 Antenna miniaturization and bandwidth enhancement using a reactive impedance substrate *IEEE Trans. Antennas Propagation* vol 52 no 9 p 2403-2414

[3] K. Sarabandi and H. Mosallaei 2006 Compact wideband UHF patch antenna on a reactive impedance substrate IEEE *Antennas Wireless Propag. Letters* vol 5 p 503-506

[4] C. Ren, L. Bernard, and R. Sauleau 2010 Investigations and design of small-size printed antennas on a reactive impedance substrate *European Conference on Antennas and Propagation* p 234-238

[5] J. S. Colbum and Y. Rahmat-Samii 1999 Patch antennas on externally perforated high dielectric constant substrates *IEEE Trans. Antennas Propagation* vol 41 no 12 p 1785-1794

[6] H. Mosallaei and K. Sarabandi 2003 A novel artificial reactive impedance surface for miniaturized wideband planar antenna design: Concept and characterization *IEEE AP-S International Symposium* vol 2 p 403-406

Numerical Simulations of Microwave Heating of Liquids: Enhancements using Krylov Subspace Methods

M R Lollchund[1], K Dookhitram[2], M S Sunhaloo[2] and R Boojhawon[3]

[1]Department of Physics, University of Mauritius, Republic of Mauritius
[2]Department of Applied Mathematical Sciences, University of Technology, Mauritius, Republic of Mauritius
[3]Department of Mathematics, University of Mauritius, Republic of Mauritius

E-mail: r.lollchund@uom.ac.mu

Abstract. In this paper, we compare the performances of three iterative solvers for large sparse linear systems arising in the numerical computations of incompressible Navier-Stokes (NS) equations. These equations are employed mainly in the simulation of microwave heating of liquids. The emphasis of this work is on the application of Krylov projection techniques such as Generalized Minimal Residual (GMRES) to solve the Pressure Poisson Equations that result from discretisation of the NS equations. The performance of the GMRES method is compared with the traditional Gauss-Seidel (GS) and point successive over relaxation (PSOR) techniques through their application to simulate the dynamics of water housed inside a vertical cylindrical vessel which is subjected to microwave radiation. It is found that as the mesh size increases, GMRES gives the fastest convergence rate in terms of computational times and number of iterations.

1. Introduction

Microwave heating of liquids is a very complex process and numerical simulations are becoming indispensable tools in helping researchers to understand the phenomenon, predict and control its behaviours. Moreover, using a numerical model, fully empirical design of microwave heating system and expensive prototype equipment for testing the design can be avoided [1]. The simulation process involves solving Maxwell's electromagnetic equations using a numerical scheme such as the Finite Difference Time Domain (FDTD) or Finite Element Method (FEM) in order to obtain the steady-state electromagnetic (EM) field distributions within the computational domain [2]. The volumetric dissipative electromagnetic power is then utilized as input to the heat transport and the Navier-Stokes (NS) equations for computations of temperature and fluid flow respectively. These equations are generally solved using a semi-implicit projection scheme for faster and reliable solutions [3]. This scheme results into a Poisson-type equation for solving the coupled pressure-velocity fields of the heated liquid. It should be noted that such type of Pressure-Poisson equation (PPE) is also obtained when other schemes such as SIMPLE, PISO or Fractional Step Method are employed to solve the NS equations. The temperature variation during the heating process can cause a change in the complex permittivity and/or permeability of the liquid, which will affect the space and time variation of the EM field [1]. Therefore, the updated EM parameters need to be re-injected into the solver for Maxwell's equations and the whole simulation process is repeated until the desired heating time or temperature is reached.

It is well known that in simulations involving heating of liquids, the major computational cost is mainly influenced by finding iterative solutions of the PPE [3]. This motivates the present work, where we aim to reduce the computational time and memory requirements of our in-house developed numerical model for microwave heating simulations by employing an appropriate solver for the PPE. A comparative computational study is carried out to determine the best performing algorithms from the existing traditional solvers Gauss Seidel (GS) and Point Successive over Relaxation (PSOR), and the Krylov Subspace method Generalized Minimal Residual (GMRES) [4], to solve the PPE. Our model is applied in the study of the dynamics of water housed inside a vertical cylindrical vessel which is subjected to microwave radiation.

2. Methodology

2.1. The Problem Studied

The problem studied in this section was first investigated in [5]. It consists of a closed vertical cylindrical vessel of radius $R = 85$ mm and height $H = 101$ mm, completely filled with water. It is assumed that microwave radiation of intensity $Q_0 = 1.125 \times 10^6$ Wm^{-3} are incident on the sides of the vessel and that the power distribution inside the vessel is radially symmetrical and is given by Lambert's law as

$$Q(r) = Q_0 \exp\left(\frac{r - R}{\delta}\right),$$

where δ is the penetration depth of the wave. The change in temperature will lead to changes in the complex permittivity of the material and hence to a change in the penetration depth. In the range 20 °C to 50 °C, the penetration depth can be approximated as a linear function of temperature as $\delta = 0.004408 + 0.0005752T$, where T is in °C and δ is in meters [5].

The characteristics of the fluid flow are obtained by solving the continuity and momentum equations which are commonly termed as Navier-Stokes (NS) equations. The heat transfer is modeled by the energy equation. The mathematical formulations of these equations are given elsewhere [6]. The no-slip boundary conditions are used for the velocity components at the side, top and bottom walls of the vessel which are also taken to be thermally insulated. The buoyancy force term in the NS equation is given by Boussinesq approximation [6]. The thermal properties of the water sample are given elsewhere [5, 6]. Initially, the temperature throughout the vessel is set at 25 °C and the liquid velocity is zero. The simulation is run for a total of 180 seconds of microwave heating.

2.2. The Pressure Poisson equation (PPE)

Upon employing a semi-implicit projection scheme and application of the above boundary conditions to the NS equations for calculating the flow fields, a Pressure-Poisson equation (PPE) is obtained. The interested reader is referred to [6] for a derivation of the PPE. Following the conventional process, the PPE can be written as the linear system

$$Ax = b, \tag{1}$$

where the coefficient matrix A is block tri-diagonal, the elements of the unknown vector x consist of the pressure values that need to be computed at each point on the grid and the elements of vector b are obtained from known terms in the NS equations [6].

2.3. The Iterative Linear Solvers

Many different iterative methods for solving equation (1) are known. In the present study we compare the performances of three well known iterative methods. The linear solvers considered are the traditional Gauss Seidel (GS) and Point Successive over Relaxation (PSOR) methods, and the Krylov projection scheme GMRES. We should mention that the GMRES method is rarely employed in the microwave heating simulation community due to its complex algorithm structure as compared to the GS and PSOR techniques [7]. Details about these solvers can be obtained from [3, 4, 6, 7]. The objective is to study their convergence rate and computational times for the problem studied using different grid sizes for different convergence tolerances. The convergence criteria for determining solution of the PPE is based on the absolute difference between two successive iterates:

$$\sum_i \left| x_i^{(n)} - x_i^{(n-1)} \right| < \epsilon,$$

where x_i is the i-th element in vector x, n is the iteration number and ϵ is the tolerance value for convergence.

3. Results and Discussions

3.1. Temperature profiles

Figure 1 compares the temperature profiles obtained inside the vessel after various heating times (60 s, 120 s and 180 s) with the work presented in [5]. The left plot of figure 1 shows the radial temperature profiles at a height of 68 mm from the bottom of the vessel and the right plot of figure 1 depicts the variation of temperature with height (measured from the bottom of the vessel) at a radius of 51 mm. Good agreement between the results are observed.

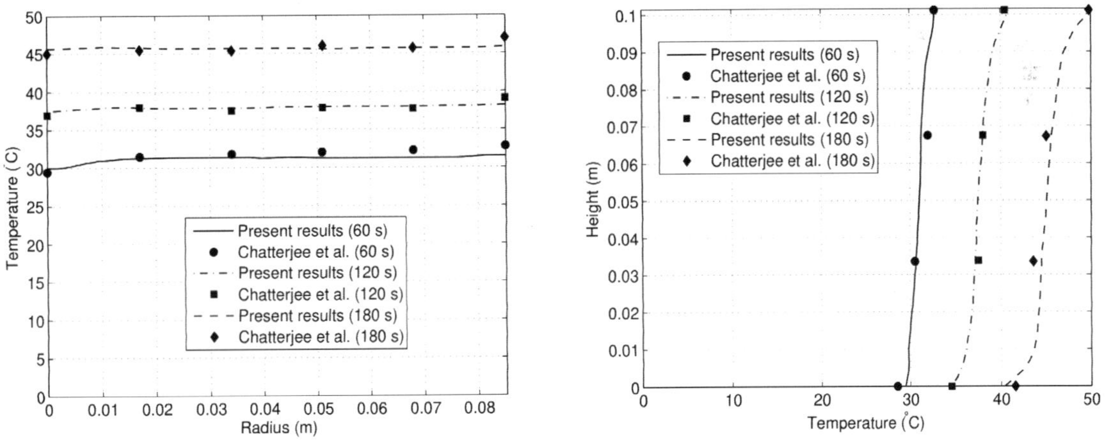

Figure 1. Comparison of temperature profiles, obtained using the present model with those published by Chatterjee et al. [5], at different times in the cylindrical vessel (Grid size: $31 \times 19 \times 62$).

3.2. Performance of the iterative linear solvers

The numerical experiments are performed for different grids. In table 1 the computational times (cpu) involved in the computation of the PPE and the corresponding number of iterations involved (n) are displayed. It should be noted that all computations were carried out on an Intel Core 2 Quad desktop computer with 2.50 GHz clock speed and 2 GB RAM. It can be observed

Table 1. Performance of three iterative schemes for solving the PPE, with different values of ϵ.

ϵ	Grid size	GS		PSOR		GMRES	
		n	cpu(s)	n	cpu(s)	n	cpu(s)
10^{-3}	$15 \times 7 \times 30$	80	0.08	76	0.07	73	0.06
	$21 \times 13 \times 42$	197	0.51	191	0.44	186	0.25
	$31 \times 19 \times 62$	423	2.52	405	2.21	395	1.07
	$41 \times 25 \times 82$	718	10.96	688	9.63	678	4.23
10^{-4}	$15 \times 7 \times 30$	356	0.19	339	0.17	331	0.11
	$21 \times 13 \times 42$	839	1.72	803	1.50	787	0.87
	$31 \times 19 \times 62$	1801	10.29	1725	9.07	1698	4.07
	$41 \times 25 \times 82$	3078	46.48	2956	40.95	2912	18.50
10^{-5}	$15 \times 7 \times 30$	936	0.43	898	0.36	885	0.18
	$21 \times 13 \times 42$	1875	3.18	1798	2.78	1771	1.32
	$31 \times 19 \times 62$	4042	23.46	3879	20.64	3821	9.11
	$41 \times 25 \times 82$	6975	110.50	6684	97.32	6589	42.81

from table 1 that the GMRES method is more efficient than both GS and PSOR schemes, especially when the domain size is increased. The number of iterations involved is nearly the same for all three methods considered but the cpu speedup for GMRES is more than twice faster than GS and PSOR in all the cases.

4. Conclusion

It was demonstrated in this work that the use of GMRES for solving the linear PPE is much more efficient than the traditional GS and PSOR schemes. In practice, one always uses a fine mesh to obtain a better accuracy of the required solution. We observed that for large grid size GMRES converges faster than the GS and PSOR methods. Moreover, the GMRES method is computationally much more efficient because it provides very accurate numerical results by using very little computational times and virtual storage than the GS and PSOR schemes. As future work, we will consider other versions of GMRES, namely the Simpler GMRES and its augmented variant [8].

References

[1] Zhao X, Yan L and Huang K 2011 *Review of Numerical Simulation of Microwave Heating Process, Advances in Induction and Microwave Heating of Mineral and Organic Materials* (Stanislaw Grundas (Ed.), ISBN: 978-953-307-522-8, InTech, DOI: 10.5772/13387)
[2] Taflove A 1999 *Advances in Computational Electrodynamics: The Finite Differences Time Domain Method* (Artech House Inc.)
[3] Johnston H and Liu J *J. Comput. Phys.* **199** 221
[4] Saad Y and Schultz M H *SIAM J. Sci. Stat. Comput.* **7** 856
[5] Chatterjee S, Basak T and Das S K *J. Food Eng.* **79** 1269
[6] Lollchund M R 2007 *Numerical Simulation of Heat and Mass transfer during Microwave Heating Processes: Study of a High Pressure Vessel* Ph.D. thesis University of Mauritius
[7] Saad Y 1996 *Iterative Methods for Sparse Linear Systems* (U.S.A: PWS Publishing Company)
[8] Boojhawon R and Bhuruth M *Future Gener. Comp. Sy.* **20** 389

IOP Conf. Series: Materials Science and Engineering **44** (2013) 012007 doi:10.1088/1757-899X/44/1/012007

Influence as a challenger to Induction for near-field wireless power transfer

H Bondar[1], S Oree[2] and K Ichikawa[3]

[1] TMMS Research Centre, Flic en Flac, Mauritius
[2] Dept. of Physics, Faculty of Science, University of Mauritius, Réduit, Mauritius
[3] Murata Manufacturing Company Limited, Kyoto, Japan

E-mail: henribondar@yahoo.fr

Abstract. We introduce what we call "Influence" WPT technology, which can be perceived as the dual of magnetic induction WPT. In Influence systems, two or more electric circuits are coupled by electric fields whereas Induction systems involve magnetic field coupling. We propose an unconventional approach to describe and study Influence, using coupling matrices, in analogy with the treatment of Inductive coupling, albeit with a larger number of parameters. In spite of their different natures, both technologies enable non-radiative near-field energy transfer at mid-range, and display interesting and sometimes counter-intuitive behaviors.

1. Introduction

Energy transfer through short distances using moving magnets or electrically charged bodies [1] had been accomplished centuries before Hertz's discovery of electromagnetic waves. Here, we provide a soft introduction to the idea of energy transfer through electric fields using the experimental arrangements of figures 1(a) and (b). Figure 1(a) shows two electrically charged distant pendulums, initially in equilibrium. If one pendulum is set into oscillatory motion, energy is coupled to the second one due to their distant coulombic force interaction, hence producing oscillations in the latter. If one of the pendulums is replaced by an elongated conductor as shown in figure 1(b), application of forced oscillations to the pendulum can induce charge oscillations along the conductor. Such electrical charge displacements caused by an external charge was historically referred to as "electrical influence" in Europe [2]. Contemporarily, the term "electric induction" is more common, but may lead to confusion.

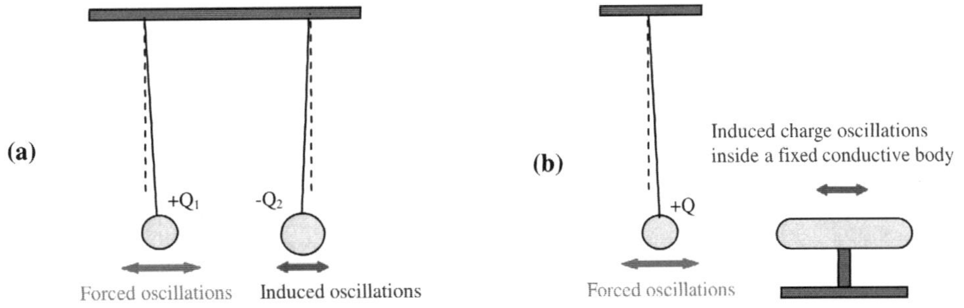

Figure 1. (a) Induced oscillations in coupled pendulums (b) Induced charge motion in a conductor.

IOP Conf. Series: Materials Science and Engineering **44** (2013) 012007 doi:10.1088/1757-899X/44/1/012007

2. A modern form of Influence

The basic concepts presented above are at the heart of modern Influence WPT devices where energy is transported between oscillating electric dipoles, just like magnetic induction WPT involves coupled oscillating magnetic dipoles (figure 2). It is well-known that the coupling between dipoles is better if they are aligned [3]. A specificity of the electrically coupled systems is that it is much easier to realize asymmetric dipoles constituted by electrode pairs (figure 3). This may bring about an increase in the coupling coefficient *k*, efficiency, range as well as other interesting and unrivalled characteristics [3].

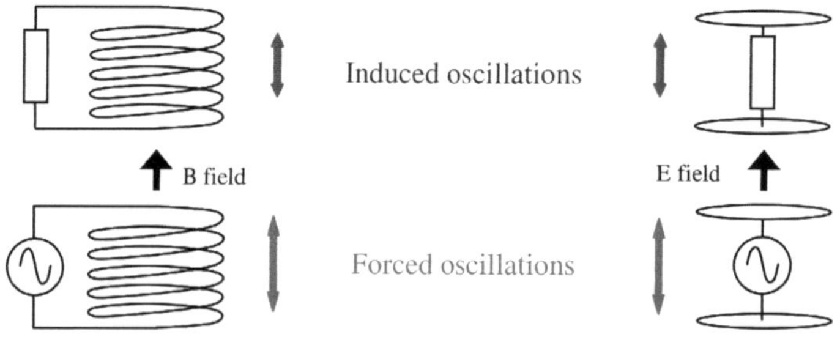

Figure 2. Magnetic (left) and electric (right) oscillating coupled dipoles in longitudinal configuration.

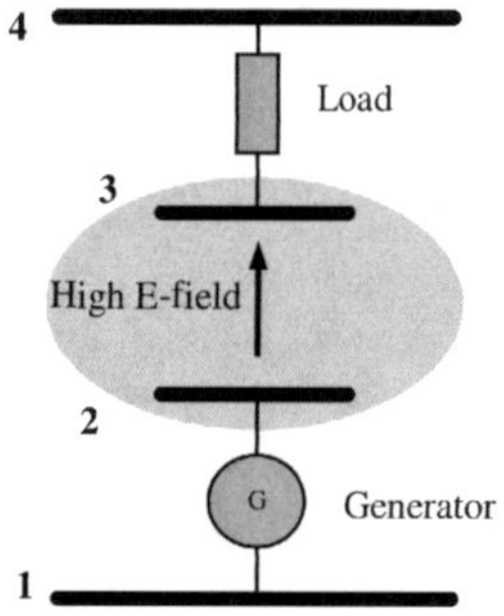

Figure 3. Longitudinal asymmetric configuration.

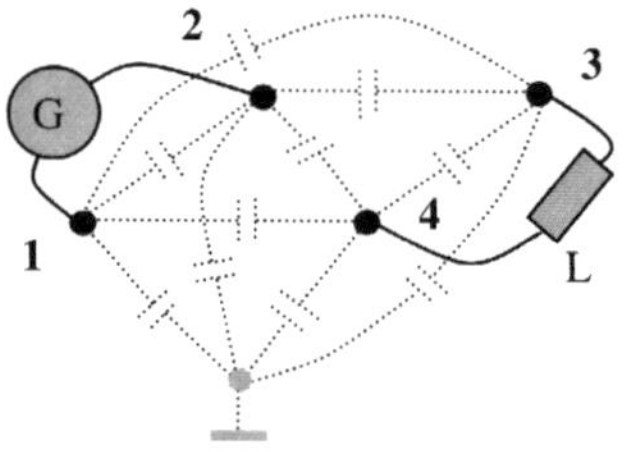

Figure 4. Equivalent Circuit representation.

Electric coupling can be represented through classical electrical circuits as in figure 4. However, in this ten-capacitor representation, some circuit nodes are virtual and some branches are crossed by virtual currents. In this work, we use an alternative representation shown in figure 5. In the circuit of figure 5, only circuit branches carrying measurable material currents are shown. We call them "open-ended representations". They enable much clearer and more realistic portrayal of devices. In contrast to Induction where only three coefficients are concerned, the full description of all the voltages involved in capacitive coupling is contained in a 4x4 matrix with at most ten independent coefficients. This gives more degrees of freedom and increased technical possibilities. For instance, using asymmetry, it is possible to concentrate the electric field in a region between small facing electrodes.

$$\begin{pmatrix} Q_1 \\ Q_2 \\ Q_3 \\ Q_4 \end{pmatrix} = \begin{pmatrix} C_1 & -C & 0 & 0 \\ -C & C_2 & 0 & 0 \\ 0 & 0 & C_3 & 0 \\ 0 & 0 & 0 & C_4 \end{pmatrix} \begin{pmatrix} V_1 \\ V_2 \\ V_3 \\ V_4 \end{pmatrix}$$

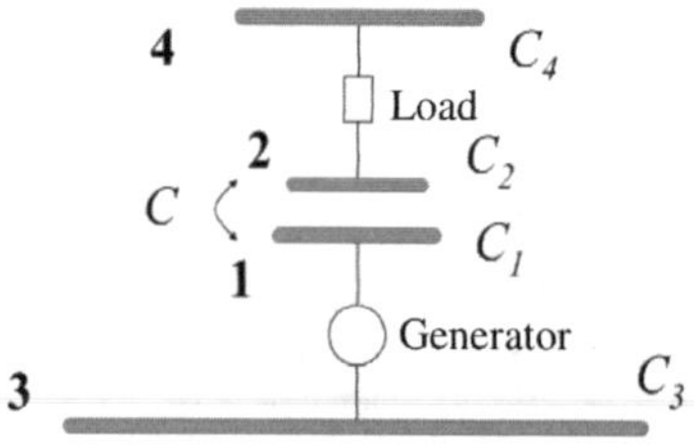

Figure 5. A possible configuration of closely spaced electrodes and its open-ended representation.

As a result, reactive energy and high voltages are concentrated between these two "active" electrodes. Furthermore the casing of the original devices can be used as the large "passive" electrodes, allowing seamless integration of the technology in existing products. The specific geometries possible for electric coupling between electrodes also lead to a higher degree of flexibility in lateral positioning compared to magnetic coupling between coils (figure 6). In particular, there is no lateral misalignment that can nullify the coupling, in contrast to coupled coils. It should be pointed out that only the active electrodes of the Influence coupled devices are shown with the passive ones assumed to be far away.

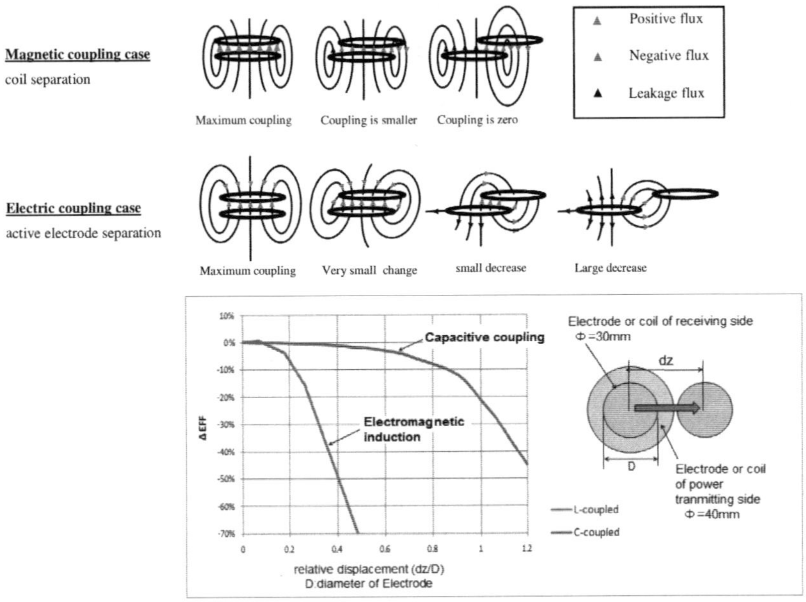

Figure 6. Efficiency fall with lateral misalignment - inductive and capacitive coupling [5].

3. Power limitations

If we base our reasoning on the everyday experience that magnetic forces between magnets can be much larger than electrostatic forces, we might hastily conclude that Influence is fundamentally limited to much lower power transfer applications than Induction. Both technologies involve large amounts of reactive energy stored in the vicinity or inside the devices. Only a very small part of this energy is typically transferred at mid-range; however the transfer occurs many times per second according to the generator frequency, and can equate to quite high levels of transmitted power.

Using the classical expression for the energy density within a region of electric field, we may define a power density according to the formula: $P_d = \omega\varepsilon E^2/2$. At 1 MHz, in dry air, we obtain a maximum reactive power level of 250 kVA/dm^3. This leaves room for large power applications even if a minute part of this reactive power is transferred as real power. Practical limitations (just like for induction systems), are associated to dissipation and efficiency rather than to electrical breakdown. For applications requiring high power or long range, EMI regulations can also introduce limitations.

4. Use of resonances

For both technologies, resonances can be used to improve efficiency [4, 5]. The main idea (exploited by Tesla himself) being to recycle the untransferred energy instead of losing it at each alternation by integrating the coupling elements within a high Q-factor RLC circuit driven at its resonance frequency. This recycling can be done on both generator and load sides (it is more efficient to recycle the nearby energy on both sides). The effect of resonance is summarized in table 1. Another important factor is that capacitors generally have higher component Q-factors than inductors, resulting in a higher efficiency or range potential for Influence systems over Induction systems.

Table 1. Summary of the effect of resonance on effective coupling.

Implementation	No resonance	Single resonance	Double resonance
Effective coupling	k	$k\sqrt{Q}$	$k\sqrt{Q_G Q_L}$

5. Impedance considerations

Exact solutions of Maxwell's equations for fields produced by elementary electric or magnetic dipoles (current loop) are available. The term "elementary" means that dipole size is small compared both to the wavelength λ (non relativistic movements) and to the measurement distance d. Generally the electric and magnetic field vectors have both a phase angle and a geometric orientation in space; their ratio cannot be characterized by a single number. However along an axis perpendicular to the dipole, electric and magnetic fields are at right angles allowing the definition of a complex impedance $Z = E/H$ which conveys information on both the amplitude ratio and the phase difference between the fields. The magnitudes of Z for electric and magnetic dipoles are given in figure 7.

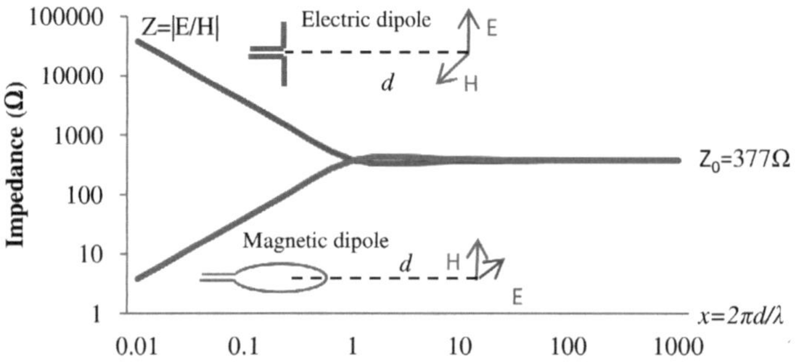

Figure 7. Impedance in the near and far fields of elementary electric and magnetic dipoles.

We note that the electric dipole's near field displays high impedance, while that of the magnetic dipole displays low impedance. As expected, they converge to the value of the wave impedance of TEM waves in the far-field. Finally, the decisive factor for Influence technologies is their high impedance character. As for high-voltage distribution lines, they need, for the same power handling, far less conductive materials than their low-voltage competitors. Moreover, the use of copper electrodes is not mandatory for Influence: they can be made of transparent conductive plastic sheets.

6. Conclusions

Power can be transferred efficiently between two distant oscillating electrical dipoles coupled through their electric fields. Electric coupling between circuits is more difficult to grasp and represent than its magnetic counterpart because virtual currents exist outside the devices. In schematics, we propose to restrict the use of the classical capacitor symbol to cases of total influence and to adopt "open-ended" circuits for more general cases. Compared to Induction systems, Influence is more tolerant to positioning and provides other flexibilities. This high impedance technology is currently being used in consumer electronics [5]. Prospection for automotive applications has started [6]. Natural extensions to large low-cost arrays of switched electrodes to cover long distances or wide surfaces are expected.

7. References

[1] Malin S and Barraclough D 2000 *Eos, Trans. Amer. Geophys. Union* **21** 233
[2] Gray J 1890 *Electrical Influence Machines* (London: Whittaker and Company)
[3] Bondar H *French Patent* 2875649 (2004)
[4] Tesla N *U.S. Patent* 645576 (1900)
[5] Goma S 2011 *Asia Electronics Industry* **182** 35-37
[6] http://techon.nikkeibp.co.jp/english/NEWS_EN/20120706/227082/, accessed 15 Oct. 2012

IOP Conf. Series: Materials Science and Engineering **44** (2013) 012008 doi:10.1088/1757-899X/44/1/012008

Experimental study of rectenna coupling at low power level

A. Douyère[1], F. Alicalapa[1], S. Rivière[2] and J.-D. Lan Sun Luk[1]

[1]Laboratoire d'Energétique, d'Electronique et Procédés, Université de La Réunion, 97715 Saint-Denis, France.

[2]OSCADI, 81 rue Georges Lebeau, 97431 Plaine des Palmistes, La Réunion.

E-mail: alexandre.douyere@univ-reunion.fr

Abstract. The experimental results presented in this paper focus on the performance of a rectenna array by studying the effect of mutual coupling between two rectennas. The measurements in several planes of the space are investigated and used to help us to define the minimum distance for future rectenna arrays that can be used at a low power density level. The single element chosen for the array is composed of a rectifier circuit and a CSPA (Circular Slot Patch Antenna). This study shows that at a distance greater than 6cm ($\lambda/2$) between two rectennas in reception, we observe that the DC received voltage is constant in the Y plane, while in the X plane, the DC received voltage remains constant whatever the distance. We deduce that these rectennas are uncoupled in this case. We can consider each rectenna like an independent system.

1. Introduction

Recently, the emergence of various wireless power technologies [1], [2] to eliminate the last cable has generated significant research interest in this area. The concept is to eliminate in the future the need for problem-prone wiring and the constraints for battery replacement. Indeed, this study is motivated by two types of applications: (a) powering of low-power sensors and (b) RF energy recycling with respect to energy consumption to natural environment. In this study we focus our attention on the power collecting system: the rectenna (the RF/DC converter) [3]. Due to the safety regulations, we consider low power level densities. But in order to increase the collected power we have to use a rectenna array. The challenge is to maximize the rectenna array power conversion efficiency for low input power density levels. To this end, we have chosen to experimentally study the effects of the mutual coupling of two rectennas in two planes of space on the RFDC conversion. The single element rectenna is composed of a compact circular slot patch antenna (CSPA), coupled with a simple and efficient rectifying circuit. The article begins with the study of the rectenna design. In this section, all the advantages to the CSPA antenna design are developed. The optimization of the rectifying circuit at f_0=2.45GHz, using a matching quadripole (four-ports) between input impedance and antenna impedance, is also presented. Finally, the overall test setup and experimental results on the mutual coupling between two rectennas are described and discussed in the subsequent sections.

IOP Conf. Series: Materials Science and Engineering **44** (2013) 012008 doi:10.1088/1757-899X/44/1/012008

2. Rectenna design

The rectifying circuit present in this section has a simple structure. This structure is also compact and efficient for low input power levels.

2.1. CSPA antenna

In this work, a circular slot patch antenna, first introduced in [4], is designed as the collecting element. Our choice was to use this type of antenna because it has many advantages in a rectenna design development. These advantages are: a poor impedance matching (which is useful for the rectenna efficiency at the different harmonics generated by the diode ($2f_0$, $3f_0$ and $4f_0$)), a good increased bandwidth, equal to 23.4%, the reduced antenna sizes, two major lobes in opposite direction, and fabrication simplicity. An illustration of the antenna is first proposed in Figure 1 and 2. This antenna realization uses the methodology as in [5]. Concerning the antenna design optimization, we use a parametrical optimization.

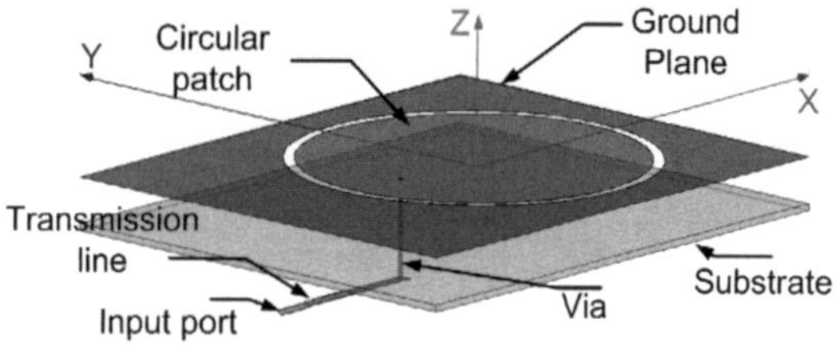

Figure 1. Illustration of the proposed CSPA.

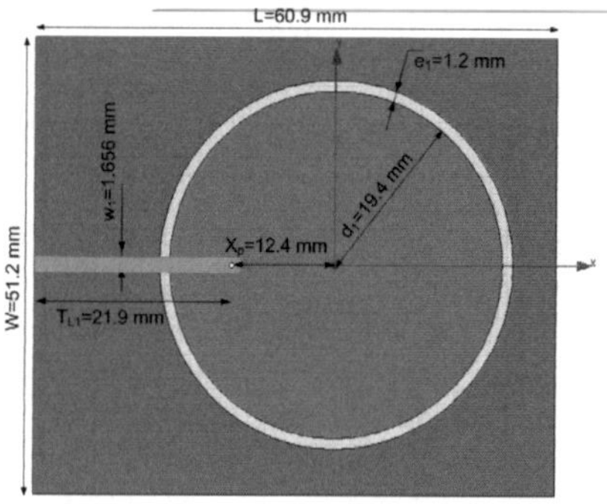

Figure 2. CSPA dimensions.

It can be noted that the antenna radiation pattern shows an advantage in the context of this study. Indeed, the radiation pattern exhibits two lobes in opposite directions (+z and z). This characteristic gives the possibility to use this antenna, in a rectenna application where the incident RF wave is composed of the direct, interfering and reflected signals. At Θ=0, the simulated antenna gain and directivity are respectively equal to 3.32dBi and 3.42dBi. At aΘ angle of 180, the gain and the directivity are respectively equal to 3.79dBi and 3.89dBi.

2.2. Rectifying circuit element design

A series configuration of the rectifying circuit was retained. It consists of a matching input circuit, followed by one Schottky diode placed in series configuration and a DC pass filter. A 12pF chip capacitor, with a quality factor Q=10 at 2.45GHz, is used to short the RF energy and pass the DC power to the resistive load. For the matching input circuit, a tuning shorted stub was used for both the diode biasing in the DC domain, and the impedance matching between the antenna and the rectifier. The rectifying circuit and the antenna are etched on ARLONs AD series substrate. The choice of the rectifier diode was focused on the SMS7630-001.

A matching circuit is essential in providing the maximum power transfer from the antenna to the rectifier circuit. As it is shown in Figure 3, to match the input impedance of diode with the CSPA antenna impedance of 50Ω, a combination of a transmission line and a shorted stub is designed. The optimization of the impedance matching (see Figure 4) consists of adjusting the length L_1 of the shorted stub and the length TL_2 of the transmission line. Given the global rectifying circuit at $P_{collected}$=-20dBm with R_{Load}=3.5kΩ, and using a Tuning method under Momentum in ADS software, the optimized length values are: L_1=2.9mm and TL_2=10.9mm.

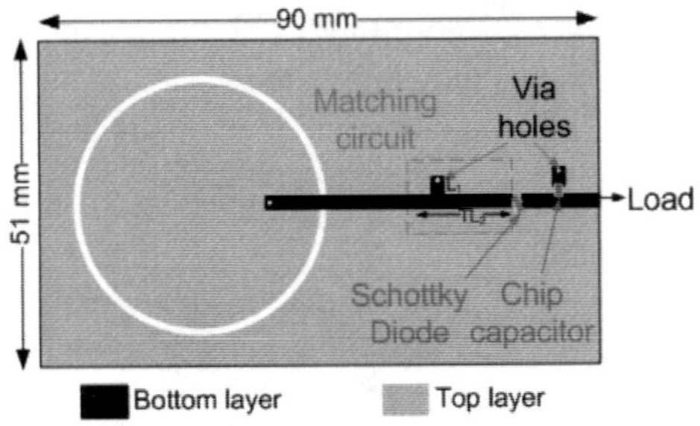

Figure 3. Layout of the rectenna.

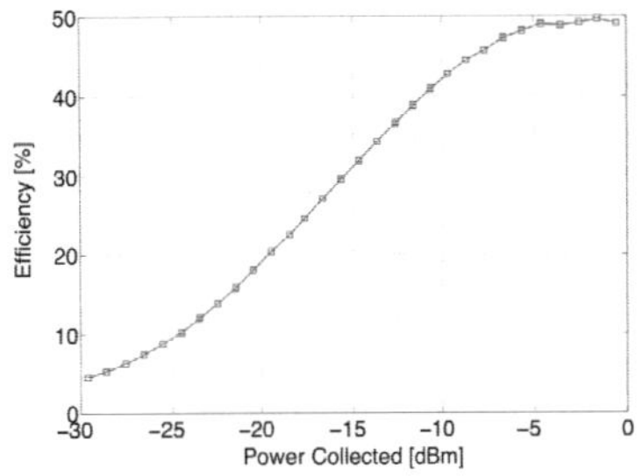

Figure 4. Measurement conversion efficiency of the rectifier versus the collected power for RL=3.5KΩ.

The width of the microstrip transmission line was chosen based on the input impedance of the antenna (50Ω) with W_1=1.656mm (see Figure 3). We measure the overall RF/DC conversion for one rectenna element. We have a good RF/DC conversion efficiency at around η=20% for a collected power of 10μW and a maximum of 50% at $P_{collected}$=0.3mW (Figure 4).

3. Measurement setup and results

After the validation of the rectifying circuit and the CSPA antenna separately, the rectenna is realized on the same unit substrate (Figure 6).

3.1. Measurement bench

We measured the effect of mutual coupling between two rectennas in the two planes of space (X and Y). The measure is performed in free space, and the experimental measurement setup is shown in Figure 5, which is automatic and controlled by the software LabView. The RF signal generator is controlled, and allows the power and frequency to be varied. The RF injected power P_{inj} into the transmitter antenna is varied from +13dBm to +28dBm in order to obtain in reception a low collected power of rectenna. Moreover, the frequency range for these measurements is 2.42Ghz to 2.47Ghz. A linearly polarized patch antenna array is used for providing the RF power to rectennas. The purpose of this study is to experimentally estimate the DC voltage received V_{DCout} by a constant resistive load of 10kΩ connected at the output of each rectenna, as a function of a distance between rectennas in each plane of space. The wireless transmission distance R is equal to 1 meter.

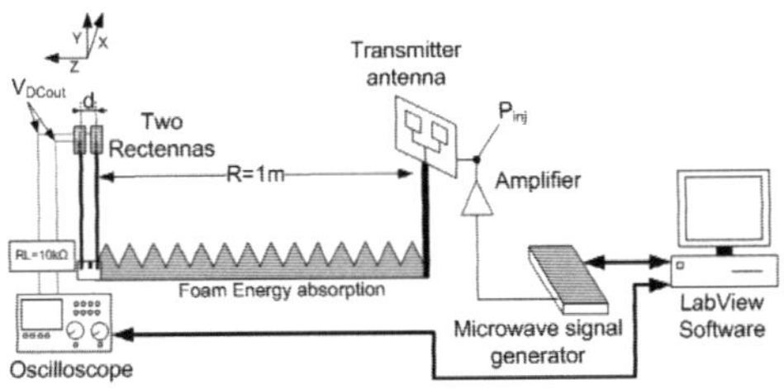

Figure 5. Experimental measurement setup.

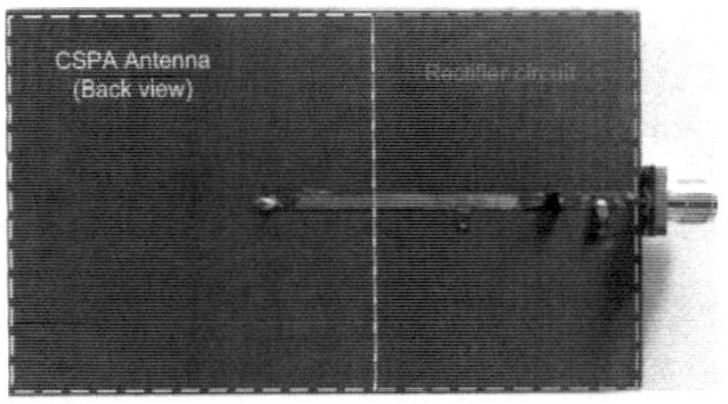

Figure 6. Photograph of the rectenna back view.

For these results, it should be noted that averaging was performed during the measurement sessions due to the high sensitivity of the rectenna elements on the measurement environment and process mismatch.

3.2. *Results*

The study begins with the measurements of the effect of the mutual coupling in the X plane of the space. For this, each element rectenna is separate from their center, in the X plane, by a distance ranging from d=5cm to d=12cm.

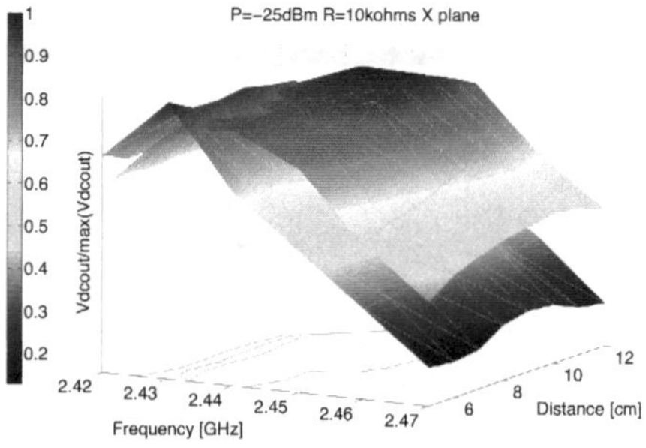

Figure 7. Coupling in the X plane at P_{inj}=+13dBm, $P_{collected}$=-18dBm.

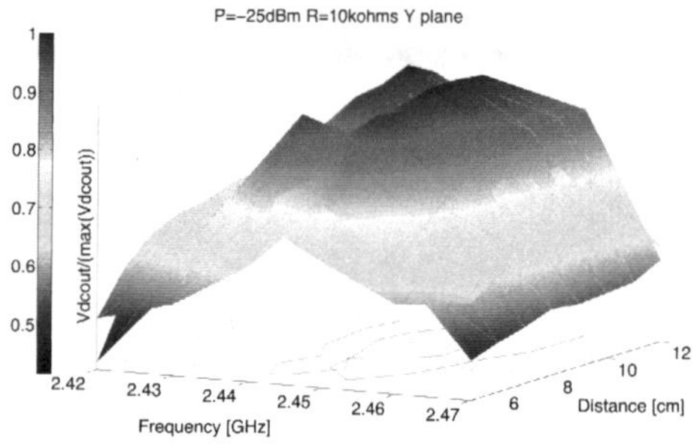

Figure 8. Coupling in the Y plane at P_{inj}=+13dBm, $P_{collected}$=-18dBm.

On the Figure 7, we can notice that the DC received voltage $V_{DCout}/(Max(V_{DCout}))$ remains almost constant as a function of the distance between the two rectennas. Thus, no significant optimal distance can be observed. On the same vein of optimizing the rectenna array power conversion efficiency for low input power levels density, the optimal distance is searched in the Y plane, as it is shown on Figure 8. In this case, we observe that a maximum DC received voltage is achieved for a distance greater than 6cm.

4. Conclusion

The rectenna structure was formed using a single diode in series configuration coupled with a CSPA antenna. The design of this antenna with a circular slot patch antenna offers several advantages. At last, good bidirectional radiation patterns with appreciable gain are obtained over the operating bands. In order to optimize a rectenna array, this experimental study shows that, in the X plane, no mutual coupling and no optimal distance significant between two rectennas are observed. In contrast, in the Y plane, a maximum DC received voltage is achieved from a distance greater than $\lambda/2$ and thus it determines a minimum distance for a maximum coupling. In view of these results, the future goal of this work will be the measurement evolution of the rectenna array efficiency in the Z plane.

References

[1] Hsuan Peng I and Chen Y-W 2012 Energy consumption bounds analysis and its applications for grid based wireless sensor networks *Journal of Network and Computer Apllications* **36** 444-51.
[2] Chen S, Yao J, and Wu Y 2012 Analysis of the Power Consumption for Wireless Sensor Network *Node Based on Zigbee International on Information and Electronics Engineering (IWIEE)* **29** 1994-98.
[3] Andia Vera G 2009 Efficient Rectenna Design for Ambient Microwave Energy Recycling *PhD Thesis University of Catalunya.*
[4] Heikkinen J, Kivikoski M 2003 A novel Dual-frequency Circularly Polarized Rectenna *IEE Antennas Wireless Propagation Letter* **2** 330-3.
[5] Rivière S, Alicalapa F, Douyère A and Lan Sun Luk J-D 2010 A compact rectenna device at low power level *Progress In Electromagnetics Research C* **16** 137-146.

Radio and Antenna Days of the Indian Ocean (RADIO 2012) IOP Publishing
IOP Conf. Series: Materials Science and Engineering **44** (2013) 012009 doi:10.1088/1757-899X/44/1/012009

The Minimization of the Extraneous Electromagnetic Fields of an Inductive Power Transfer System

James McLean and Robert Sutton
TDK R&D Corp., Cedar Park, Texas, USA
E-mail: jmclean@tdkrf.com

Abstract. The efficiency of inductive wireless power transfer (IPT) systems has been extensively studied. However, the electromagnetic compatibility of such systems is at least as important as the efficiency and has received much less attention. We consider the net magnetic dipole moment of the system as a figure of merit. That is, we seek to minimize the magnitude of the net dipole moment in order to minimize both the near magnetic fields and the radiated power. A 20 kHz, 3.3 kW, IPT system, representative of typical wireless vehicular battery charging systems, is considered and it is seen that one particular value of load impedance minimizes the net dipole moment while another, distinct, value maximizes efficiency. Thus, efficiency must be traded off, at least to some extent, in order to minimize extraneous electromagnetic fields.

1. Introduction

Inductive power transfer (IPT) is the most common approach for high-power wireless power transfer applications, such as the recharging of batteries for electric vehicles [1,2]. In keeping with vehicular power requirements and providing reasonably short recharging times, such systems range in power from approximately 3 to 20 kW for consumer applications and up to 150 kW for commercial vehicles [3]. A representative system is shown in Fig. 1. This configuration employs what has become known as "circular" couplers [4]. Many IPT systems with operating frequencies ranging from 10 kHz to 150 kHz and all with very high efficiency have been successfully demonstrated. Therefore, selection of operating frequency will more likely be motivated by electromagnetic compatibility requirements [5]. From ref. [3] it can be deduced that load powers on the order of 3-20 kW and separation distances of 200 mm will be common in IPT systems existing in consumer environments. This then sets constraints on the minimum extraneous electromagnetic field that can be achieved through careful design.

2. Loosely-Coupled Transformer Model

An inductive power transfer system functions fundamentally as a loosely-coupled transformer. While some electromagnetic radiation necessarily takes place, it causes only a minor perturbation in the primary and secondary currents because of the minuscule electrical size of such a system. A system operating at 150 kHz ($\lambda \approx 2$ km) (probably the highest frequency under consideration for IPT) with a separation of 200 mm and a lateral dimension of 700 mm (comparable to the size of the systems in ref. [4]) still qualifies as quasi-magnetostatic. Thus, the currents can be accurately determined using network theory. Once the primary current I_1 and secondary

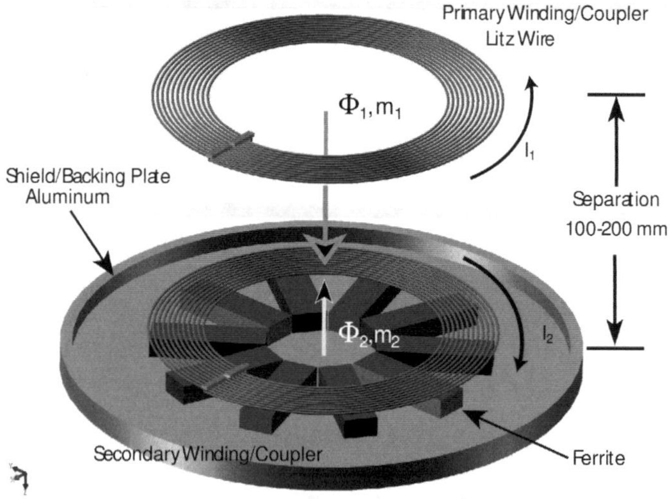

Figure 1. Circular couplers for IPT. Note that both the primary and secondary couplers are wound with the same sense. Magnetic material for top coupler not shown. Ideally, the magnetic fluxes Φ_1 and Φ_2 and the dipole moments m_1 and m_2 are in phase and equal in magnitude to provide cancelation.

current I_2 in Fig. 1 are known, both the near magnetic field and the far radiated electromagnetic field can be determined from the current distribution using electromagnetic field theory. A loosely-coupled transformer will generate an electromagnetic field which in the most general representation is an infinite series of spherical harmonics. The lowest-order terms are three orthogonal magnetic dipoles. Thus, in the lowest order, both the near-field magnetic field and the radiated electromagnetic field are largely proportional to the net dipole moment [6], $\vec{m}_{net}$. This vector phasor quantity is the sum of the dipole moments of the primary and secondary windings, $\vec{m}_1$ and $\vec{m}_2$ respectively. Since in the ideal case, the axes of the couplers are parallel, we consider the scalar phasor quantity $m_{net} = m_1 - m_2$, for the current directions indicated. This quantity, in turn, is proportional to the difference of the currents, $I_1 - I_2$ when the couplers are identical and their directions are defined as in Fig. 1.

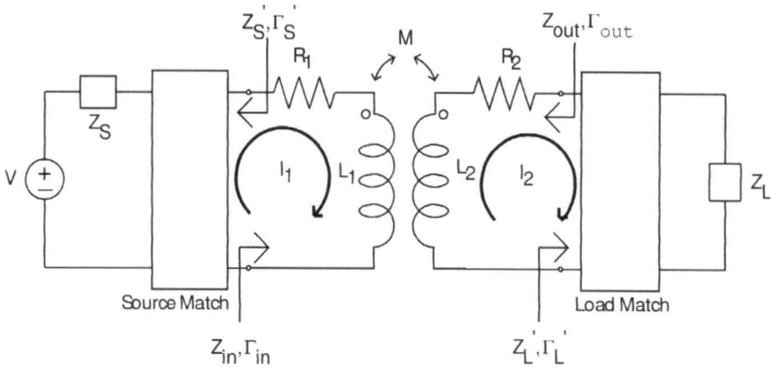

Figure 2. Coupled inductor circuit representation for a loosely-coupled transformer including near-field inductive power transfer systems. Note that the arrangement of dots corresponds to Fig. 1; if $I_1 = I_2$, the net dipole moment is zero.

For such a pair of coupled inductors connected to a complex load as shown in Fig. 2, the secondary current I_2 is related to the primary current I_1 as:

$$I_2 = \frac{j\omega M}{j\omega L_2 + R_2 + Z_L'} I_1, \tag{1}$$

where R_1 and R_2 represent the losses in the primary and secondary windings respectively. Note that in Eqn. 1 the matching networks are absorbed in Z'_s and Z'_L. For a lossless, tightly-coupled transformer, the primary and secondary fluxes essentially cancel if $|j\omega L_2| >> |Z'_L|$. For loosely-coupled inductors with unity turns ratio, flux cancelation would still occur if $j\omega M = j\omega L_2 + R_2 + Z'_L L$. This would require a negative load resistance for perfect cancelation. However, it can be gleaned that the net minimum dipole moment still requires:

$$jX'_L \approx j\omega\left(M - L_2\right).$$

If $R_2 = 0$, a purely reactive load $jX'_L = j\omega\left(M - L_2\right)$ can make the primary current I_1 and secondary current I_2 equal and hence the dipole moments of the primary and secondary equal in magnitude and opposite in phase thereby completely canceling the net dipole moment. This is equivalent to connecting the two windings in series-opposing fashion, a similar arrangement to an MRI z-gradient coil. This situation is the extreme case of no transferred power and no power dissipated in the secondary winding. On the other hand, in the absence of losses, the denominator in Eqn. 1 could be forced to zero with $jX'_L = -j\omega L_2$, although this results in an infinite primary-side input impedance. Yet another value of reactance, $jX'_L = -j\omega\left(L_2 + M\right)$ would cause $I_1 = -I_2$ and thus providing direct addition of the dipole moments. It provides the same magnetic field as would be obtained connecting the two coils in series-aiding fashion in a similar arrangement to a Helmholtz coil.

However, it is clear from the above equation that the net moment cannot be zero as long as a resistive component of the load exists, even if the secondary winding is lossless. When finite power is transferred to the load, the dipole moments cannot cancel completely, but rather a finite minimum net dipole moment exists. Thus, if we set the load power, $P_L = |I_2|^2 \operatorname{Re}\left(Z'_L\right)$, we can choose $I_2 = \sqrt{\frac{P_L}{\operatorname{Re}\left(Z'_L\right)}}$, setting the phase of I_2 to zero. Thus, we consider the quantity:

$$\sqrt{\frac{P_L}{\operatorname{Re}\left(Z'_L\right)}}\left[\frac{I_1}{I_2} - 1\right] = \sqrt{\frac{P_L}{\operatorname{Re}\left(Z'_L\right)}}\left[\frac{j\omega L_2 + R_2 + Z'_L}{j\omega M} - 1\right] \propto m_1 - m_2. \tag{2}$$

This has a minimum value of $\frac{2\sqrt{R_2}}{\omega M}\sqrt{P_L}$ when $Z'_L = j\omega M - j\omega L_2 + R_2$. Thus, we consider:

$$\frac{|m_{net}|}{\sqrt{P_L}}\frac{1}{\frac{2\sqrt{R_2}}{\omega M}} = \; = \frac{\omega M}{2\sqrt{R_2}}\frac{1}{\sqrt{Re\left(Z'_L\right)}}\left|\frac{j\omega L_2 + R_2 + Z'_L}{j\omega M} - 1\right|.$$

This gives the magnitude of the net dipole moment relative to the minimum value that can be obtained with fixed power delivered to the load .

3. Maximum Efficiency versus Net Dipole Moment

The following circuit parameters are representative of a 3.3 kW IPT system for vehicular battery charging: $f = 20$ kHz, $L_1 = L_2 = 540$ μH, $M = 275$ μH, $k = .501$, $R_1 = R_2 = R_s = 271$ mΩ. For these values, the efficiency of the system, $\eta = \frac{P_{load}}{P_{in}}$ is plotted as a function of the real and imaginary components of the load impedance in Fig. 3. The net dipole moment is plotted in Fig. 4. The maximum efficiency is obtained with $Z'_L = \sqrt{R_S^2 + \omega^2 M^2} - j\omega\, L_2$ while the minimum dipole moment is obtained with $Z'_L = j\omega M - j\omega L_2 + R_s$. The efficiency when the load is fixed for minimum dipole moment is 31.3 %. On the other hand, the magnitude of the net dipole moment at the maximum efficiency load it about 8 times that of its minimum value.

4. Conclusions

It is not possible to simultaneously minimize the net dipole moment while maximizing efficiency in an IPT system. Thus, a compromise between these two goals must be sought.

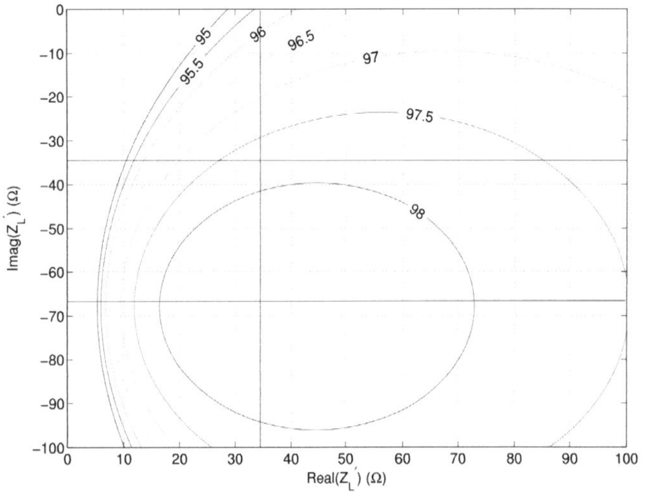

Figure 3. Efficiency plotted as a function of the the real and imaginary components of the load. The value of Z'_L for maximum efficiency is:
$Z'_L = 34.55 - j67.\ \Omega$.
The efficiency is 98.44 % at this point.

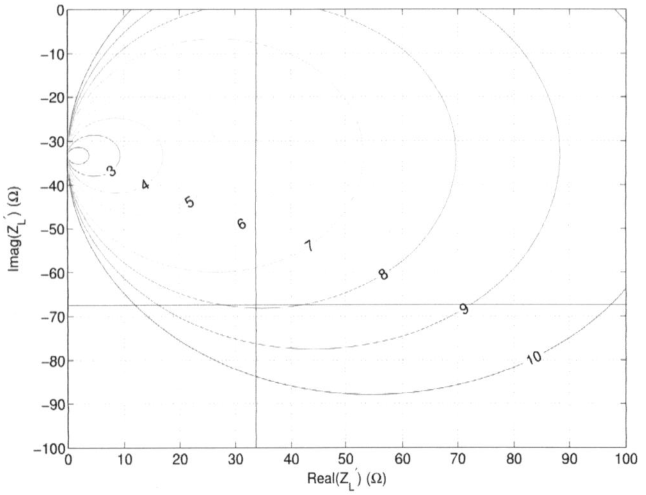

Figure 4. Net dipole moment plotted as a function of the the real and imaginary components of the load. The value of Z'_L for minimum dipole moment is:
$Z'_L = .271 - j33.3\ \Omega$.
The contour plot is normalized to the dipole moment at this value of effective load impedance.

References

[1] Grant Covic and John Boys, "Inductive Power Transfer (IPT) Powering Our Future," The University of Auckland, New Zealand, Power Electronics Research Group, avail at http://ceme.ece.illinois.edu/seminars/CEME910Covic.pdf 2010.

[2] Grant Covic and John Boys, "Possibilities in Inductive Energy Transfer," Power Electronics Research Group, The University of Auckland, New Zealand, 2010.

[3] G. Covic and I. Teerlinck, "SAE International J2954 Wireless Charging Task Force Report from: Interoperability Subteam", March 12, 2012.

[4] M. Budhia, G. A. Covic,and J. T. Boys, "Design and Optimization of Circular Magnetic Structures for Lumped Inductive Power Transfer Systems," *IEEE Trans. Power Electronics*, Vol. 26, No. 11, pp. 3096-3108, Nov. 2011

[5] J. McLean and R. Sutton, "The Electromagnetic Compatibility of Wireless Inductive Automotive Battery Chargers and LF Animal RFID Tags" in *Proc. Asia Pacific International Symposium on Electromagnetic Compatibility*, pp. 921-924, Singapore, 21-24 May, 2012.

[6] J. R. Bergervoet and H. Van Veen, "A large-loop antenna for magnetic field measurements", in *Proc. Zurich Int. Symp. (EMC-Zurich 1989)*, Zürich, Switzerland, pp. 2934, Jan., 1989.

Electromagnetic modeling and definition of antenna specifications and positions for radio system deployment in confined environments

J. Avella Castiblanco[1], D. Seetharamdoo[1], M. Berbineau[1], M. Ney[2], F. Gallee[2]

[1] IFSTTAR, LEOST, 20 Rue Elise Reclus - BP 70317 - 59666 Villeneuve D'Ascq Cedex, France.

[2] Telecom Bretagne, Dept. Microondes, Brest, France.

E-mail: `jorge.castiblanco@ifsttar.fr`, `divitha.seetharamdoo@ifsttar.fr`

Abstract. In this paper, the definition of antenna specifications for radio system deployment as well as a methodology to integrate them in confined environments is presented. Modal-based analysis and optimization techniques for mode-weight adjustments are used to determine correct co-excitation of the propagating modes and fulfill the desired specifications.

1. Introduction

Wireless communication system design in confined environments is complex due to harder electromagnetic and spatial constraints. Indeed the number of on-board wireless systems is increasing and polymers are increasingly used to build modern, lighter railway vehicles. The investigation on novel antennas is of outstanding importance to guarantee reliable communication for systems such as control-command. Appropriate methodologies for antenna design and integration have to be developed to face the challenges on such platforms.

Research work on antenna integration in railway environments by using metamaterials is performed in the framework of the METAPHORT project initiated by the French National Institute of Science and Technology for Transport (IFSTTAR). The aim of this project is to develop metamaterial antennas and to establish a methodology for the antenna integration in urban underground railways. Train-to-Wayside communication systems for control command applications for railways in an urban underground environment based on 802.11 a and 802.11 g WLAN standards are considered. Antenna integration in this environment requires appropriate numerical models of the electromagnetic wave propagation accounting for multi-scale geometries such as tunnels and objects, and in a reasonable computation time. Appropriate methodologies for the antenna integration based on these models will be proposed and validated.

The paper is organized as follows. Section 2 presents a methodology for the correct antenna integration. Results using our methodology is shown in section 3. Finally, conclusions and on-going work are developed.

2. Methodology for Meta-material antenna design

Metamaterials can be described as artificial media tailored to control wave propagation. They can provide a higher degree of freedom in the design of antennas for wireless systems tailored for specific environments by controlling the EM energy stored or radiated around the antenna. For the development of novel antenna solutions in tunnels different inputs are required as depicted in Fig. 1.

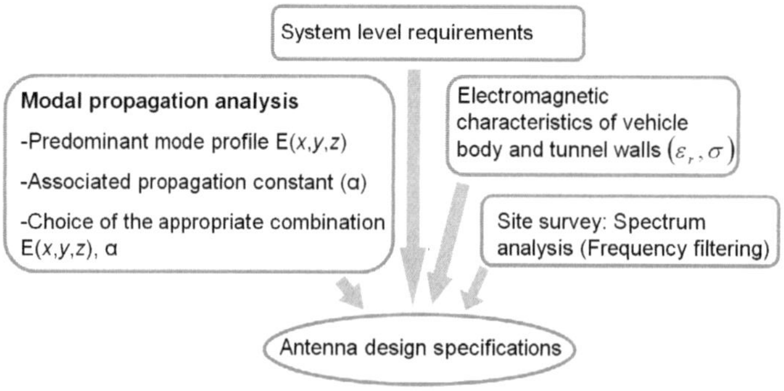

Figure 1. Novel metamaterial antenna design methodology.

2.1. Antenna design for tunnels

The specifications and parameters involved in the antenna design for confined environments, such as tunnels, have not yet been established in the scientific or technical literature. A few considerations on the obstruction, separation, installation and maintenance can be found in the U. K. guidance on train rooftop antenna positioning (GK/GN06602) as well as by some manufacturers. However, the problem of antenna operation in tunnel environments is not addressed. Thus, a methodology for antenna design is needed and appropriate models providing physical insight must be considered. Various studies have been done considering several models to analyze radiation characteristics and positioning of antennas in tunnels [1] or simply to model the wave-propagation in these environments [2]. These models are based on asymptotic approaches and they thus fail for antennas which are strongly affected by the surrounding environment. A novel methodology for finding appropriate field specifications and optimum antenna positioning can be obtained by using a modal approach with a full-wave method, such as the Transmission Line Matrix method (TLM) [3].

2.2. Methodology for finding antenna specifications

An efficient modal approach for the analysis of guiding structures is applied to find the fields [4]. Different optimum positions and types of excitation are involved due to multi-modal propagation in these environments. These modes may be combined in some way to produce a desired effect, *e.g.*, the maximization of the power transferred by the modes.The particularity of this approach is that only the cross-section of the tunnel is considered while modeling the field propagation in the longitudinal direction. Tunnels with arbitrary cross section, loaded or unloaded, can be analyzed with this approach.

2.2.1. Description of the Methodology The proposed methodology is divided into five steps. The first one concerns the calculation of the characteristic mode parameters α, β, $\widehat{\mathbf{E}}(x,y)$ and $\widehat{\mathbf{H}}(x,y)$. Second, the modes are discriminated by their power carried through the tunnel so that only modes with the highest power are considered. Then, an optimization procedure of the transmitted power on the weight coefficients is carried out. A constrained solution in terms of the maximum or minimum power density that can be transmitted or received, respectively, must be found. In the fourth step, the optimum weight coefficients to maximize the mean power along the propagation direction are obtained. Finally, the optimum fields specifications, type of excitation and source positioning are determined by using the optimum weights. A detailed description of this methodology can be found in [3].

3. Results

The wave propagation in a rectangular tunnel with transverse dimension $w = 7\text{m} \times h = 5\text{m}$ was considered. The tunnel walls were modeled by a lossy dielectric with $\epsilon_r = 12$ and $\sigma = 0.02\text{Sm}^{-1}$. The power density specifications were defined to be 0 dB Wm^{-2} at $z_{Tx} = 0$ and -40 dBWm^{-2} at $z_{Rx} = 1,000$ m. This tunnel was simulated using our modeling approach. The attenuation, phase constants, and the field profiles of the E^y_{11} and E^x_{11} modes were determined and compared with Marcatilli's approximate theory, results are shown in Fig. 2. The attenuation and dispersion

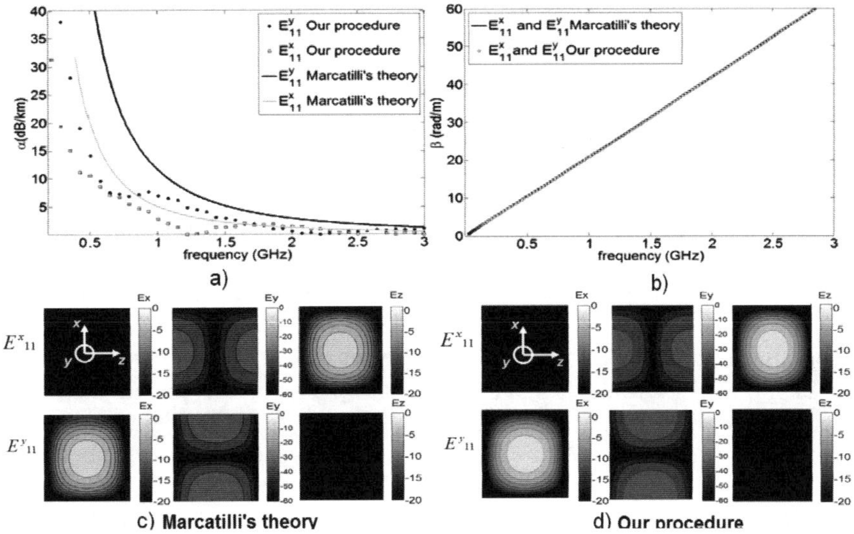

Figure 2. Attenuation, dispersion and field profiles for the E^y_{11} and E^x_{11} modes.

characteristics calculated by our modeling approach are in good agreement with Marcatilli's theory. However, the attenuation constants obtained for low frequencies are different most probably due to the fact that in Marcatilli's approach the fields at the corners are canceled unlike in our numerical method. The optimum electric fields to excite the rectangular tunnel are then calculated by using the above results from the modal approach; optimum weight coefficients are then defined using our methodology. Optimum positions where the mean power along the longitudinal distance is maximum were also calculated. By way of illustration, the three field components of the electric field that satisfy the power density specifications in Tx were calculated at $z = 0$, as shown in Fig. 3.

Intuitively, one could expect that optimum fields are most influenced by the lowest attenuated modes. This result can be confirmed by observing that the points where the x-component of the electric field in Fig. 3 is minimum at corners, and the z-component is minimum at half of the width as it occurs for the E^x_{11} mode. This observation can also be confirmed as the x-component is maximum at one-third, as well as one-quarter of the width as for both E^x_{21} and E^x_{31} modes.

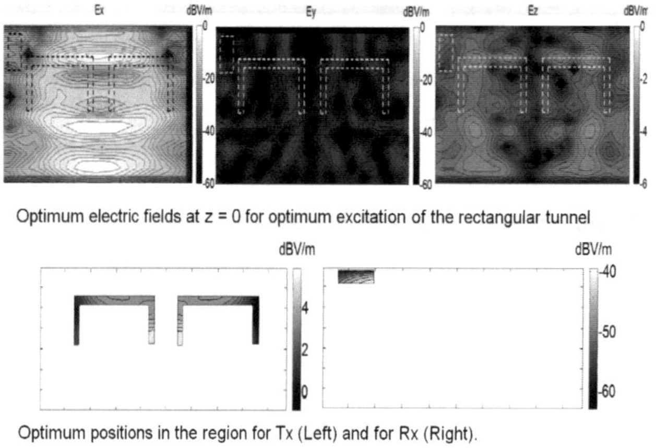

Figure 3. Modeling approach.

Optimum positions to locate the transmitter are those where the mean power along the longitudinal distance is maximum, i.e., for the points close to the white regions in Fig. 3.

4. Conclusions

A methodology for the definition of antenna specifications for radio system deployment and antenna integration in confined environments has been presented. The methodology relies on a numerical model based on modal analysis. Then, optimization techniques were applied, and mode-weight adjustments are used to determine correct co-excitation of the propagating modes and ensure a good power transfer from the transmitter to the receiver. Results are presented for a tunnel in terms of complex propagation constant and mode profiles. Good agreement with an approximate model has been shown, thus validating our numerical results. The antenna design methodology for positioning and determination of specifications was then applied. It should be noted that this methodology has to be repeated for all the critical scenarios varying for traffic considerations, different tunnel cross sections, multiple tracks and different distances between the transmitter and receiver.

Acknowledgments

This work was funded by the French National Agency within the ANR-VTT 2009 program in the framework of the METAPHORT project. The authors also acknowledge support from the I-TRANS railway cluster of the Nord Region of France, as well as the Images et Reseaux cluster of the Brittany region.

5. References

[1] Y. Huo, Z. Xu, H. dang Zheng, and X. Zhou, "Effect of antenna on propagation characteristics of electromagnetic waves in tunnel environments," in *Microelectronics Electronics, 2009. PrimeAsia 2009. Asia Pacific Conference on Postgraduate Research in*, Jan. 2009, pp. 268 -271.

[2] D. Didascalou, "Ray-optical wave propagation modelling in arbitrarily shaped tunnels," Ph.D. dissertation, Institut fur Hochstfrequenztechnik und Elektronik der Universitat Karlsruhe (TH), 2000.

[3] J. Avella Castiblanco, D. Seetharamdoo, M. Berbineau, M. Ney, and F. Gallee, *Wave propagation. Chapter: "Electromagnetic wave propagation modeling for finding antenna specifications and positions in tunnels of arbitrary cross-section"*., ser. ISBN 980-953-307-493-6. INTECH, 2012.

[4] ——, "Surface boundary conditions for lossy dielectrics to model electromagnetic wave propagation in tunnels," *European Conference on Antennas and Propagation (EUCAP)*, pp. 2331 – 2335, April 2011.

Analysis of a subwavelength Z-shaped metamaterial

A Dhouibi[1], S N Burokur[2,3,4], A de Lustrac[2,3] and A Priou[1]
[1] LEME, Univ. Paris-Ouest, EA 4416, 92410 Ville d'Avray, France
[2] IEF, Univ. Paris-Sud, CNRS, UMR 8622, 91405 Orsay cedex, France
[3] Univ. Paris-Ouest, 92410 Ville d'Avray, France

E-mail: shah-nawaz.burokur@u-psud.fr

Abstract. A Z-shaped easy made planar meta-atom is presented as an alternative design to the conventional electric-LC (ELC) resonator for achieving negative permittivity. Transforming the LC topology of the resonator helps to facilitate transposition of geometrical parameters for the optical regime and also to improve the metamaterial homogeneity. Our approach aims to simplify the resonator design and achieve a lower resonance frequency without being limited by fabrication techniques. The electromagnetic behaviour is investigated through both simulations and experiments in the microwave regime. Our results show that the developed meta-atom exhibits a strong electric response to normally incident radiation and can be used very effectively in producing materials with negative permittivity. The Z-shaped structure is also shown to present a higher effective medium ratio and can possibly find various applications in planar high frequency passive circuits due to the simplicity of fabrication.

1. Introduction

Metamaterials have gained lots of attention within the scientific community over the past decade [1-3] because of their capabilities which go beyond conventional materials [4] and because of their applications in novel class coordinate transformation based devices such as invisibility cloaks [5-9], rotators [10], retroreflectors [11], Luneburg lenses [12], waveguide tapers [13] and directive antennas [14-16]. A typical metamaterial is an artificially engineered structure made of a periodic array of subwavelength metallic or dielectric inclusions called meta-atoms. These inclusions can be presented as LC resonant elements with the inductance and capacitance defined by the shape and geometrical dimensions of the latter resonator. The resonators can exhibit permittivity and/or permeability values in the effective medium regime where the ratio between the operating wavelength and the unit cell size λ/a must be greater than 4 [17]. With typical designs incorporating resonant and dispersive elements much smaller than the operating wavelength, a homogenization scheme is possible, which leads to meaningful interpretation of the effective permittivity ε, permeability μ and index n.

In this paper, we propose a practical way to change the LC equivalent circuit topology of the conventional ELC resonator [18] in the effective medium regime. It is reported that our transformation can lead to a better metamaterial homogeneity; in other words we can observe resonance at a lower frequency for similar geometrical dimensions. We investigate numerically and experimentally the electromagnetic behaviour of the Z-shaped structure evolved from the transformation and negative real parts of ε are demonstrated in the microwave domain.

[4] To whom any correspondence should be addressed.

2. Design of the Z-shaped meta-atom

We first consider the unit cell of the ELC resonator introduced in Ref. 18 and presented in figure 1. This resonator has both inductive and capacitive elements, where only the capacitive element couples strongly to the electric field. By continuous transformation, eliminating the *y*-oriented lateral wires, it can be reduced to an I-shaped resonator with a capacitive gap referred to as Igap resonator. We shall note that by removing these two side arms, we reduce the inductance of the element. On the counterpart, we add a lateral gap on both sides of the initial middle one. However, the parallel capacitances created by the two lateral gaps are supposed to be negligible compared to the middle one, which suggests that the Igap resonator will show a higher resonant frequency than the ELC one. Furthermore, without changing the lattice constant we compensate a part of the reduced inductance by transforming the Igap into a Zgap resonator. Adding physical length when passing from Igap to Zgap structure produces a higher phase delay and increases the total inductance of the metamaterial without requiring additional fabrication steps. Due to this increase in inductance, the Zgap resonator will present a resonance at a lower frequency. To facilitate the metamaterial fabrication process, we further transform the Zgap resonator into a Z-shaped one by removing the capacitive gap.

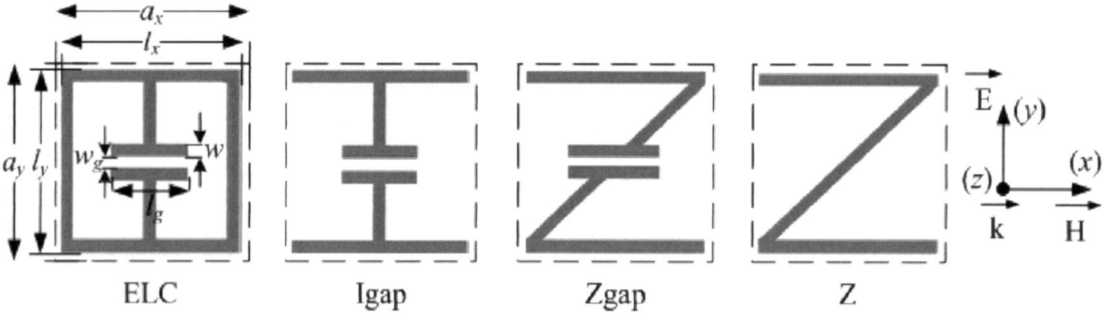

Figure 1. Design of the Z-shaped resonator from the ELC one. The geometrical dimensions are: $a_x = a_y = 6$ mm, $l_x = 5.7$ mm, $l_y = 5.8$ mm, $w = 0.3$ mm, $w_g = 0.45$ mm and $l_g = 2$ mm.

3. Simulations and experiments

The properties of the structures are characterized numerically using a Finite Integral Time Domain (FITD) Maxwell's equations solver, Microwave Studio Suite by CST. A single unit cell is simulated together with necessary periodic boundary conditions. Floquet mode ports are used to launch an incident plane wave on the different resonators. The dielectric spacer used throughout this study is single face copper-cladded epoxy with a relative dielectric constant of 3.9, a tangential loss of 0.02 and a thickness $t_s = 0.4$ mm. Total sample size is 18 × 18 cells on a 120 mm × 120 mm dielectric board. For an electromagnetic wave incident with a wave vector and field polarization of figure 1, the different resonators will exhibit an electric resonance.

Calculated reflection and transmission spectra of a single layer structure are presented in figure 2. Microwave measurements have been done on fabricated prototypes in an anechoic chamber using an Agilent 8722ES network analyzer and two standard FLANN® 2-18 GHz wideband band horn antennas serving as the source and the receiver. Phase referencing and normalization have been performed in transmission and reflection. In the transmission measurements, the microwaves are incident normal to the sample surface. Transmission measurements are calibrated to the transmission between the horns with the sample removed. The reflection measurements are done by placing the source and receiving horns on the same side of the sample and bouncing the microwave signal off the sample. The source and receiver horns are each inclined with an angle of about 4-5° with respect to normal on the sample surface. The reflection measurement is calibrated using a sample-sized sheet of copper as a reflecting mirror. Measured reflection and transmission coefficients are compared to the simulated ones. There is a very good qualitative agreement between simulations and measurements.

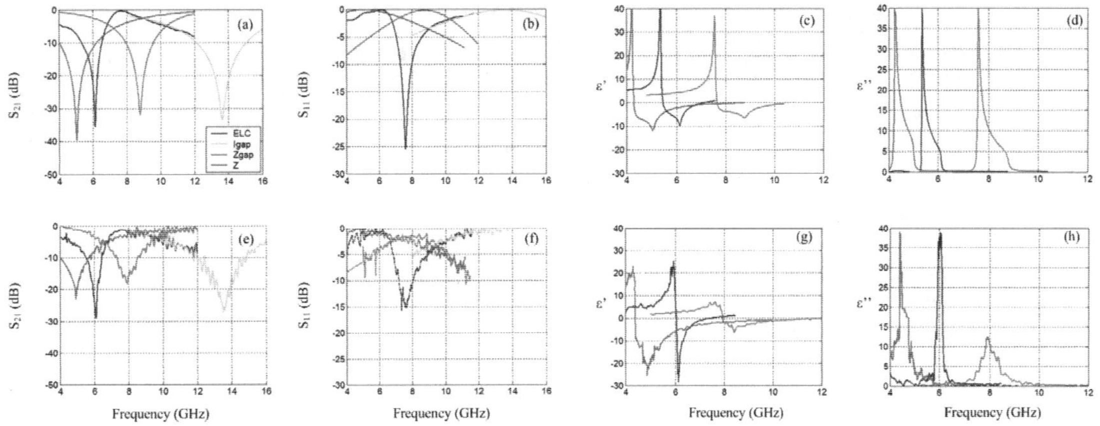

Figure 2. Computed and measured characteristics of the different resonators.

Calculated and measured magnitudes of S_{21} show clearly a resonance dip for each type of resonator. The ripples that appear in the measured responses are due to noise in the signals. They can be eliminated by a time domain gating of the frequency domain *S*-parameters. The Igap resonator presents a resonance frequency at 13.6 GHz, compared to 6.0 GHz for ELC. This shift toward higher frequencies was predictable since the inductance is greatly reduced when transforming ELC into Igap. To decrease the resonance frequency of Igap, we therefore increase its inductance by using slant lines on each side of the gap instead of vertical ones, therefore converting it to Zgap. This procedure causes a frequency downward shift of approximately 33%, which suggests that the capacitance created by the gap acts mainly to drive the LC resonance and that the LC resonance frequency is mainly based on the inductance of the structure. When cancelling the middle capacitive gap to design the Z-shaped resonator, the resonance frequency is considerably decreased. This is because two capacitances are connected in series in the Zgap structure. One is created between two consecutive cells due to the periodicity of the structure and another one is formed by the gap in the middle of the Z. In the Z-shaped structure, only the capacitance between two consecutive cells is present. According to the dimensions w_g and l_g used for the gap, the capacitance of the latter gap is almost half of the one created by the periodicity. Thus the resonance frequency of the Z is found to be 1.8 times smaller than that of the Zgap, if we consider similar inductance in both structures. And finally, in terms of effective medium ratio λ/a, an improvement of 1.9 can be observed for the Z-shaped resonator compared to the conventional ELC.

4. Effective parameters

Using reflection and transmission responses from a single layer of the sample, effective parameters can be extracted using the retrieval procedure described in [19]. This is possible since the structure period along the propagation direction is very small compared to the working wavelength. In performing the retrieval, we assume a *z*-direction size $a_z = 6$ mm corresponding to a cubic unit cell as proposed in [18]. Extracted permittivity ε and permeability μ are shown for ELC, Zgap and Z resonators in the various parts of figure 2. Concerning Igap resonator, parameters extraction is not possible since the ratio between the operating wavelength and the unit cell size, i.e. the effective medium ratio λ/a is low (~3.7). As stated above, the resonators present an electric response but no magnetic response, i.e., the real part of μ is close to unity over the frequencies of interest. However, the retrieved parameters do not strictly comply with this principle. The real part of μ becomes anti-resonant. In fact, this anomaly is an artefact introduced during parameter extraction in a frequency range where the medium is highly spatially dispersive [18]. From figure 2, we observe a strong electric resonance, with the real part of the permittivity ranging from positive to negative values in the vicinity of the resonance. The values issued from measurement data agree very well with those calculated from numerical simulations. It must be noted that the extraction from the measured spectra are presented without any fitting. From the simulated and measured real parts of ε, it can also be noted that both Z-shaped meta-atoms (with and without gap) present a wider negative permittivity frequency band compared to the ELC resonator. The proposed Z-shaped resonator can be very useful in applications where permittivity gradient is desired, such as in devices designed from transformation optics concept [5-16].

5. Conclusion

In summary, we have presented an experimental demonstration of a negative permittivity from a Z-shaped meta-atom in the microwave domain. This proposed electric resonator allows to considerably improve the effective medium ratio λ/a observed from a conventional electric LC (ELC) resonator. Due to the simplicity of the structure and the absence of capacitive gap, a transposition of geometrical parameters can be considered for the optical regime

6. Acknowledgements

This work was supported by the EADS Company Foundation through the METAQOPT project, contract No. 090-AO09-1006. A. Dhouibi acknowledges support for his PhD scholarship from EADS Company Foundation.

References

[1] Shelby R A, Smith D R and Schultz S 2001 *Science* **292** 77
[2] Yen T J, Padilla W J, Fang N, Vier D C, Smith D R, Pendry J B, Basov D N and Zhang X 2004 *Science* **303** 1494
[3] Smith D R, Pendry J B and Wiltshire M C K 2004 *Science* **305** 788
[4] Veselago V G 1968 *Sov. Phys. Usp.* **10** 509
[5] Leonhardt U 2006 *Science* **312** 1777
[6] Pendry J B, Schurig D and Smith D R 2006 *Science* **312** 1780
[7] Schurig D, Mock J J, Justice B J, Cummer S A, Pendry J B, Starr A F and Smith D R 2006 *Science* **314** 977
[8] Cai W, Chettiar U K, Kildishev A V and Shalaev V M 2007 *Nat. Photonics* **1** 224
[9] Kanté B, de Lustrac A, Lourtioz J M and Burokur S 2008 *Opt. Express* **16** 9191
[10] Chen H, Hou B, Chen S, Ao X, Wen W and Chan CT 2009 *Phys. Rev. Lett.* **102** 183903
[11] Ma Y G, Ong C K, Tyc T and Leonhardt U 2009 *Nat. Mater.* **8** 639
[12] Kundtz N and Smith D R 2010 *Nat. Mater.* **9** 129
[13] Ma H F and Cui T J 2010 *Nat. Commun.* **1** 21
[14] Tichit P H, Burokur S N and de Lustrac A 2009 *J. Appl. Phys.* **105** 104912
[15] Tichit P H, Burokur S N, Germain D and de Lustrac A 2011 *Phys. Rev. B* **83** 155108
[16] Tichit P H, Burokur S N, Germain D and de Lustrac A 2011 *Elec. Lett.* **47** 580
[17] Koschny T, Markos P, Economou E N, Smith D R, Vier D C and Soukoulis C M 2005 *Phys. Rev. B* **71** 245105
[18] Schurig D, Mock J J and Smith D R 2006 *Appl. Phys. Lett.* **88** 041109
[19] Smith D R, Schultz S, Markos P and Soukoulis C M 2002 *Phys. Rev. B.* **65** 195104

IOP Conf. Series: Materials Science and Engineering **44** (2013) 012012 doi:10.1088/1757-899X/44/1/012012

New trends in antenna design: transformation optics approach

P H Tichit[1], **S N Burokur**[1,2,3] **and A de Lustrac**[1,2]
[1] IEF, Univ. Paris-Sud, CNRS, UMR 8622, 91405 Orsay cedex, France
[2] Univ. Paris-Ouest, 92410 Ville d'Avray, France

E-mail: shah-nawaz.burokur@u-psud.fr

Abstract. Transformation optics is an emerging field offering a powerful and unprecedented ability to manipulate and control electromagnetic waves. Using this tool, we demonstrate the design of novel antenna concepts by tailoring their radiation properties. The wave manipulation is enabled through the use of engineered dispersive composite metamaterials that realize a space coordinate transformation. Numerical simulations together with experimental measurements are performed in order to validate the coordinate transformation concept. Near-field cartography and far-field pattern measurements performed on fabricated prototypes agree qualitatively with Finite Element Method (FEM) simulations. It is shown that a particular radiation pattern can be tailored at ease into a desired one by modifying the electromagnetic properties of the space around the radiating element. This idea opens the way to novel antenna design techniques for various application domains such as aeronautical and transport fields.

1. Introduction

Transformation optics or transformation electromagnetics (also called coordinate transformation) is a powerful mathematical tool that is used to generate a new transformed space from an initial one where solutions of Maxwell's equations are known by manipulating electromagnetic waves. As a first step, it consists in imagining a virtual space with desired topological properties, which will contain the underlying physics. This approach has been revived when J. B. Pendry et al. [1] have proposed an interpretation where permeability and permittivity tensors components can be viewed as a material in the original space. It is as if the new material mimicks the defined topological space. Since this pioneering work of J. B. Pendry and that of U. Leonhardt et al. [2], transformation optics is an emerging field where Maxwell's equations are form invariant under a coordinate transformation. It offers an unconventional strategy to the design of novel class metamaterial devices. The most striking application conceived so far via coordinate transformation concept is the invisibility cloak [3]. Concerning antenna applications, focusing lens antennas [4-6] and the engineering of radiation patterns [7] have been proposed. The performances of an omnidirectional retroreflector [8] and Luneberg lenses [9] have also been experimentally demonstrated. An octave-bandwidth horn antenna has been experimentally validated for satellite communications [10]. Recently, techniques of source transformation [11-13] have offered new opportunities for the design of active devices with source distribution included in the transformed space.

[3] To whom any correspondence should be addressed.

Using this last approach, we review the design of three antennas where the radiation pattern is tailored specifically in each case. The first one concerns an ultra-directive antenna obtained by stretching a source into an extended coherent radiator [14-16], which is implemented through the use of judiciously engineered metamaterials and the device is shown experimentally to produce an ultra-directive emission. The idea has been extended to a second device, a wave bending one [17], so as to achieve a steered beam antenna via a rotational coordinate transformation. Experimental measurements have shown a beam steering as much as 66°. Finally, we present the numerical design of a quasi-isotropic antenna achieved by expanding the space around a directive source [18].

2. Ultra-directive antenna

The ultra-directive antenna is based on the transformation of a cylindrical space into a rectangular one. The schematic principle of the transformation is presented in figure 1(a). The theoretical underlying physics of the transformation involved here has been detailed recently in [14]. The concept is as follows: the imagined space of our proposed antenna is obtained by transforming a flat isotropic cylindrical half-space with zero Riemann curvature tensor described in polar coordinates $\{r, \theta\}$ into a flat space in squeezed Cartesian coordinates. x', y' and z' are the coordinates in the virtual transformed rectangular space and x, y, z are those in the initial real cylindrical space. We assume free space in the cylinder, with isotropic permeability and permittivity tensors ε_0 and μ_0. In [15], we have shown that the coordinate transformation can be implemented by a material described with $e = 0.15$ m and $L = 0.05$ m by $\mu_{xx} = 1$, $\mu_{yy} = \frac{1}{(\varepsilon_{xx})^2}$, and $\varepsilon_{zz} = 4(\varepsilon_{xx})^2$.

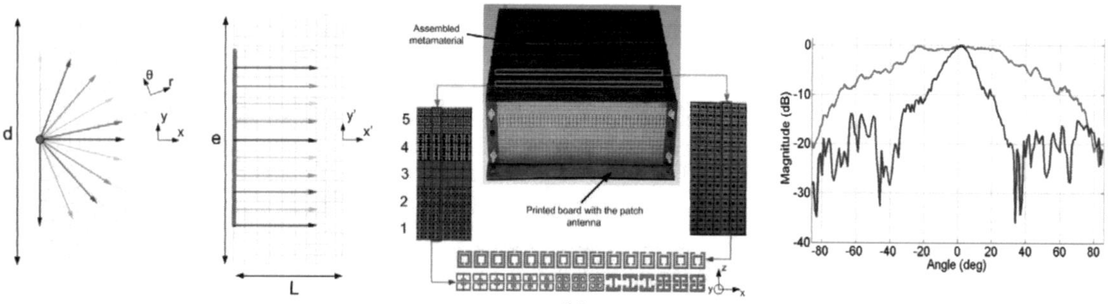

(a) (b) (c)

Figure 1. (a) Transformation of a cylindrical space into a rectangular one. (b) Photography of the antenna. The inserts show the permittivity (left) and the permeability (right) layers of the material. (c) Measured E-plane radiation patterns of the source (red) and metamaterial antenna (blue).

Discrete values are then created for the desired variation of μ_{yy} and ε_{zz} to secure a practical realization producing experimental performances close to theory. Figure 1(b) shows the photography of the fabricated prototype. A microstrip square patch antenna is used as radiating source. A surrounding material made of alternating electric metamaterial and magnetic metamaterial layers is used to capture the emanating omnidirectional radiation from the patch source and transform it into a directive one. The metamaterial is a discrete structure composed of five different regions where permittivity and permeability vary according to the above equation.

The axial permittivity ε_{zz} and permeability μ_{yy} show respectively values ranging from 0.12 to 4.15 and from 1.58 to 15.3. The bulk metamaterial is assembled using 56 layers of dielectric boards on which subwavelength resonant structures are printed. 28 layers contain SRRs and 28 others contain ELCs [15]. Each layer is made of 5 regions of metamaterials corresponding to the discretized values. Far-field patterns measurements are performed in an anechoic chamber, where the metamaterial-based antenna is used as emitter and a wideband [2 GHz – 18 GHz] is used as receiver. The E-plane radiation pattern is measured at 10.6 GHz. Figure 1(c) presents the comparison between simulations and experiments for the patch source alone and the metamaterial antenna. The transformation of the patch's omnidirectional radiation into a directive is clearly established. A narrow half-power beamwidth of 13° is observed for the measured antenna.

3. Steered beam antenna

We consider a source radiating in a rectangular space. Theoretically this radiation emitted from the latter source can be transformed into an azimuthal one using transformation optics. The transformation procedure is noted F(x',y') and consists in bending the emission. Figure 2(a) shows the operating principle of this rotational coordinate transformation. The coordinate transformation is implemented by a material defined by $\varepsilon_{rr} = \left(\frac{1}{br}\right)^2 \Big/ 1.7$, $\varepsilon_{\theta\theta} = 2.8$ and $\mu_{zz} = 1.7$.

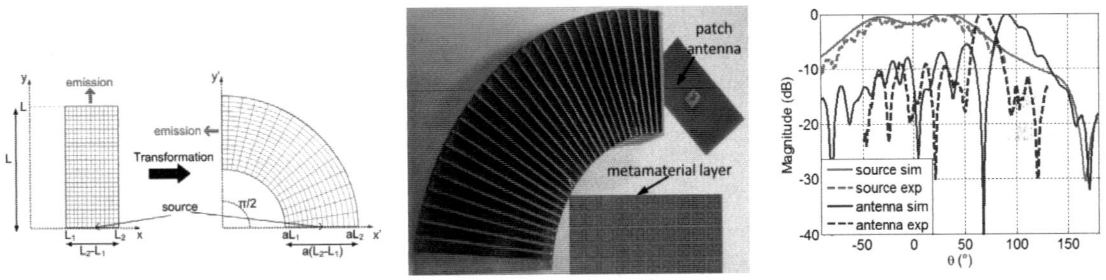

(a) (b) (c)

Figure 2. (a) Schematic principle of the 2D rotational coordinate transformation. The emission in a rectangular space is transformed into an azimuthal one. (b) Photography of the antenna. (c) Simulated and measured E-plane radiation patterns of the source (red) and metamaterial antenna (blue).

To validate experimentally the azimuthal directive emission, the device shown in figure 2(b) is fabricated. A microstrip square patch antenna is used as radiating source. The metamaterial is a discrete structure composed of 10 different regions where permittivity and permeability vary. The bulk metamaterial is assembled using 30 layers of dielectric boards on which subwavelength resonant structures are printed. The layers are mounted 1 by 1 in a molded matrix with a constant angle of 3° between each. A commercially available liquid resin is used as host and is closely linked to $\varepsilon_{\theta\theta}$. Its measured permittivity is close to 2.8. From the radiation patterns in figure 2(c), we can clearly observe the transformation of the omni-directional far-field radiation of the patch antenna into a directive one which is further bent at an angle of 66°.

4. Isotropic antenna

In this section, coordinate transformation is applied to transform directive emissions into isotropic ones. An intuitive schematic principle to illustrate the proposed method is presented in figure 3(a). Let us consider a source radiating in a circular space and a circular region bounded by the blue circle around this source limits the radiation zone. The "space stretching" coordinate transformation consists in stretching exponentially the central zone of this delimited circular region represented by the red circle. The material parameters required for this transformation are: $\mu_{rr} = 1$, $\mu_{\theta\theta} = \left(\frac{r'}{qr(r'-\alpha)}\right)^2$ and $\varepsilon_{zz} = \left(\frac{r}{r'}\right)^2$.

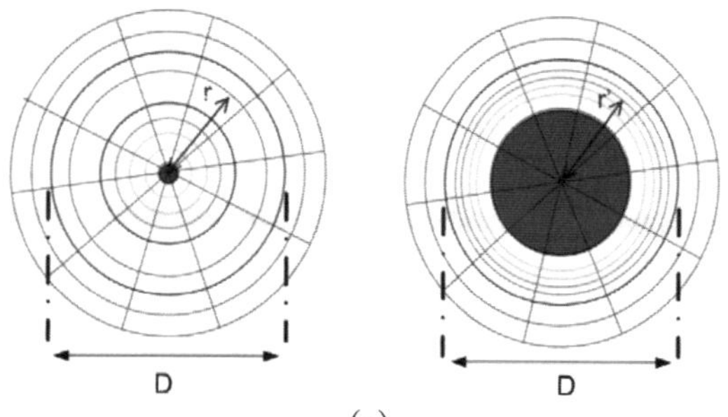

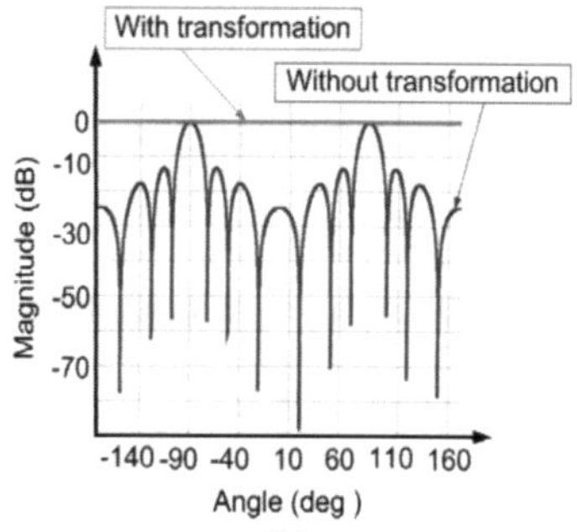

Figure 3. (a) Schematic principle of the space stretching coordinate transformation. (b) Simulated E-plane radiation patterns of the source (blue) and metamaterial antenna (red).

Numerical simulations show that a quasi-perfect isotropic emission can be effectively achieved when a high value is used for the expansion factor q.

5. Conclusion

To summarize, we have shown how coordinate transformation can be applied to manipulate and control electromagnetic waves at will in order to tailor desired radiation patterns in the antenna fields. Experimental measurements performed on the directive and beam steerable antennas have shown concluding results, making these devices compatible for aeronautical and transport domains.

References

[1] Pendry J B, Schurig D and Smith D R 2006 *Science* **312** 1780
[2] Leonhardt U 2006 *Science* **312** 1777
[3] Schurig D, Mock J J, Justice B J, Cummer S A, Pendry J B, Starr A F and Smith D R 2006 *Science* **314** 977
[4] Jiang W X, Cui T J, Ma H F, Yang X M and Cheng Q 2008 *Appl. Phys. Lett.* **93** 221906
[5] Mei Z L, Bai J, Niu T M and Cui T J 2010 *PIER M* **13** 261
[6] Jiang Z H, Gregory M D and Werner D H 2011 *Phys. Rev. B* **84** 165111
[7] Garcia-meca C, Martinez A and Leonhardt U 2011 *Opt. Express* **19** 23743
[8] Ma Y G, Ong C K, Tyc T and Leonhardt U 2009 *Nat. Mater.* **8** 639
[9] Kundtz N and Smith D R 2010 *Nat. Mater.* **9** 129
[10] Lier E, Werner D H, Scarborough C P, Wu Q and Bossard J A 2011 *Nat. Mater.* **10** 216
[11] Luo Y, Zhang J, Ran L, Chen H and Kong J A 2008 *PIERS Online* **4** 795
[12] Allen J, Kundtz N, Roberts D A, Cummer S A and Smith D R 2009 *Appl. Phys. Lett.* **94** 194101
[13] Popa B I, Allen J and Cummer S A 2009 *Appl. Phys. Lett.* **94** 244102
[14] Tichit P H, Burokur S N and de Lustrac A 2009 *J. Appl. Phys.* **105** 104912
[15] Tichit P H, Burokur S N, Germain D and de Lustrac A 2011 *Phys. Rev. B* **83** 155108
[16] Tichit P H, Burokur S N, Germain D and de Lustrac A 2011 *Elec. Lett.* **47** 580
[17] Wu X, Tichit P H, Burokur S N, Kirouane S, Sellier A and de Lustrac A 2012 *Microwave Opt. Technol. Lett.* **54** 2536
[18] Tichit P H, Burokur S N and de Lustrac A 2011 *Opt. Express* **19** 20551

Metamaterial-based Fabry-Pérot leaky wave antennas: low profile, high directivity, frequency agility and beam steering

S N Burokur[1,2,3] **and A de Lustrac**[1,2]
[1] IEF, Univ. Paris-Sud, CNRS, UMR 8622, 91405 Orsay cedex, France
[2] Univ. Paris-Ouest, 92410 Ville d'Avray, France

E-mail: shah-nawaz.burokur@u-psud.fr

Abstract. The analysis and design of subwavelength metamaterial-based Fabry-Pérot (FP) leaky wave antennas (LWAs) are presented. The antennas under investigation are formed by embedding a feeding source in a cavity composed of a Perfect Electrical Conductor (PEC) surface and a metasurface reflector. Several configurations of such antennas are presented to achieve different desired performances such as: high directivity, frequency agility and beam steering.

1. Introduction

According to the earlier work of Trentini [1], the thickness h of a FP cavity antenna is given by the resonance condition:

$$h + d\sqrt{\varepsilon_r} = (\phi_{PRS} + \phi_r)\frac{\lambda}{4\pi} \pm N\frac{\lambda}{2} \tag{1}$$

where ϕ_{PRS} is the reflection phase of the PRS reflector, ϕ_r is the reflection phase of the reflector near the antenna, and N is an integer corresponding to the order of the cavity's electromagnetic mode. d and ε_r are respectively the thickness and the relative permittivity of the antenna's dielectric board. Recent developments on FP LWAs have shown the possibility to design subwavelength resonant cavities by minimizing the total reflection phase shift $\phi_{PRS} + \phi_r$ [2-5]. Indeed, Artificial Magnetic Conductor (AMC) surfaces and metasurfaces have been proposed for applications as Partially Reflective Surface (PRS) in the design of compact cavity antennas. These surfaces present a tailored reflection phase characteristic varying from +180° to −180° and crossing 0° at the resonance frequency.

In this present work, we report the design of metamaterial-based resonant cavities for different applications. Firstly, we will show how low-profile configuration is achieved and also how directivity and gain can be considerably increased. And secondly, reconfigurability in terms of frequency and beam angle will be presented.

2. Low-profile highly directive FP LWA

FP cavity antennas are formed by placing a feeding source between two reflectors as shown in figure 1(a). In general, one of the two reflectors is PEC surface so as to prevent undesired backward radiation. The second reflector must present a high reflectivity in order to act as the so-called Partially

[3] To whom any correspondence should be addressed.

Reflective Surface (PRS). This PRS is used to confine the electromagnetic waves in the cavity. Our recent works [4, 5] have shown that metasurfaces composed simultaneously of an inductive and capacitive grid are suitable for this purpose. The inductive grid and capacitive grid are respectively made of a periodic array of copper mesh and square patches. The grids are patterned on the two faces of a dielectric substrate as illustrated by the unit cell in figure 1(b). This metasurface presents an LC resonance where the reflection phase varies from +180° to -180° and crosses 0° at the resonance as shown by the S-parameters in figures 1(c) and 1(d). As it can be seen in equation (1), the reflection phase values of the PRS (ϕ_{PRS}) are needed to calculate the thickness h of the cavity versus the frequency at which maximum boresight directivity can be obtained. Subwavelength thicknesses (h < λ/10) can therefore be expected at frequencies above resonance.

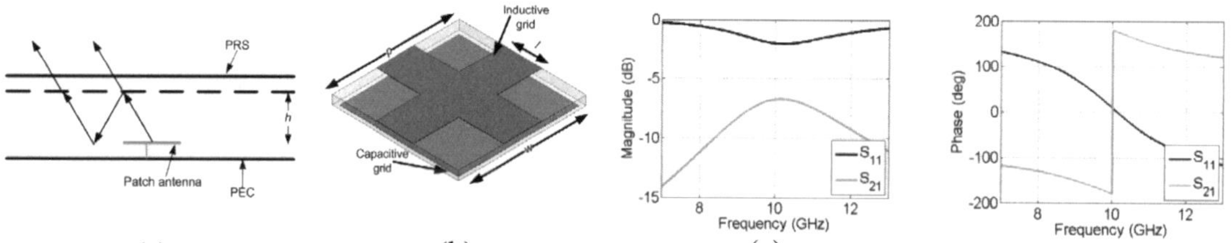

(a) (b) (c) (d)

Figure 1. (a) Schematic of a metamaterial-based FP LWA. (b) Unit cell of the PRS. (c) Reflection and transmission magnitudes of the PRS cell. (d) Reflection and transmission phases of the PRS cell.

Figure 2(a) shows, for example, the E- and H-plane radiation patterns of a λ/30 thick cavity antenna with lateral dimensions 10 cm x 10 cm designed for a 9.7 GHz operation [4, 5]. The antenna directivity is found to be equal to 19 dBi. In order to reach higher directivity, multiple primary sources can be used to feed the cavity antenna. As it can be observed in figure 2(b), the inter-element spacing of the array plays an important role on the enhancement of the directivity. This effect is highlighted in table 1 where the performances of cavities for the different inter-element spacing are presented [6, 7].

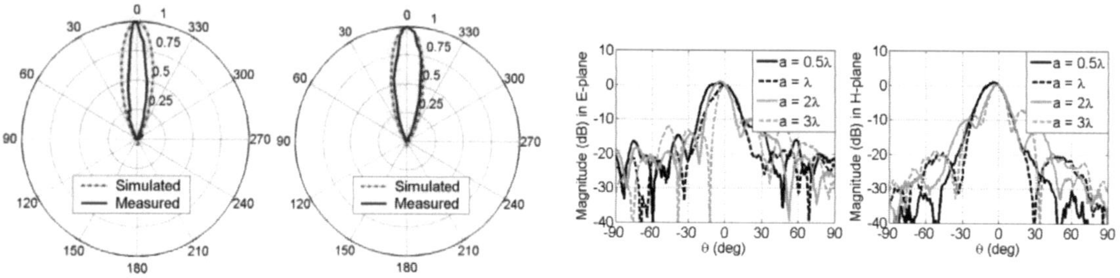

(a) (b)

Figure 2. E- and H- plane radiation patterns of FP cavity antenna fed by (a) a single source, (b) multiple sources.

For a = 0.5λ, a measured directivity of 19 dBi is obtained at 8.93 GHz. This value is very close to that of a cavity fed by a single source (see for e.g. [4, 5]). So, it is worth to note that conversely to classical antenna arrays, the directivity is not doubled each time that the number of sources is doubled. For a = λ, a measured directivity of 20.9 dBi is noted at 9.07 GHz, showing clearly an enhancement of 1.9 dB with regard to the case a = 0.5λ. It is also very important to note that the sidelobes level of the patch array is considerably reduced when embedded in the cavity. 23.21 dBi and 25.35 dBi is respectively deduced from the measured planes for a = 2λ and a = 3λ. When the case a = 3λ is compared to a = 0.5λ, an increase of 6.35 dB is obtained for the directivity, which is comparable to an increase from a single patch element to a 2 x 2 patch array. The measured sidelobes level are higher (~ –8dB in the H-plane) for the case a = 3λ. However, this sidelobes level is still low compared to the sidelobes level of the source alone. It is well known that an inter-element spacing of an array higher than λ leads to high sidelobes level and also to the apparition of grating lobes. The increase in directivity is due to the fact that a larger surface of the PRS is illuminated when a is bigger.

Table 1. Performances of the cavity antennas with h = 1.5 mm and for different values of inter-element spacing a.

a (mm)	Resonance frequency (GHz)	Maximum directivity (dBi)	Secondary lobes level (dB)
0.5λ	9.13	19 @ 8.93 GHz	-12
λ	9.37	20.9 @ 9.07 GHz	-19
2λ	9.18	23.21 @ 8.94 GHz	-10
3λ	9.21	25.35 @ 8.96 GHz	-8

3. Beam steering in FP LWA

By analogy with conventional antennas where several radiating elements are placed in array with a phase shift between each element in order to have a beam pointing in an off-normal direction, the same principle can be applied to FP cavity antennas. To achieve beam steering, phase shifts can be applied between the cells of the PRS. For example, the gap spacing g can be varied along the PRS [8, 9], as shown in figure 3(a). Figure 3(b) shows the beam deflection in the E-plane radiation pattern for different values of δg. To achieve electronic beam steering, varactor diodes are inserted in the cells and by applying different bias voltages along the surface, a phase variation is created [9]. Figure 3(c) shows the beam deflection in the E-plane for different values of δV.

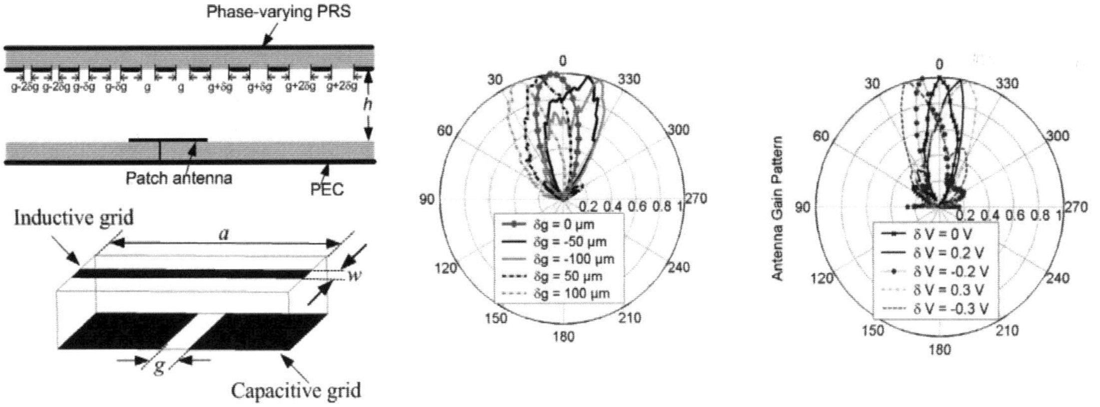

(a) (b) (c)

Figure 3. (a) Schematic view of a cavity composed of a phase variable PRS. (b) Passive beam steering. (c) Active beam steering.

4. Frequency agile FP LWA

Conversely to beam steerable cavity antennas, we do not need a locally phase-varying PRS for frequency agility applications. What we seek is the ability to change the resonance frequency of the PRS and this is possible by changing simultaneously and in the same manner the capacitance value of the varactor diodes [10-12]. An electronically controlled PRS, similar to the schematic principle of figure 3(a) is designed to operate near 2 GHz in base station antennas for mobile phone communication systems. The primary source of the cavity is a wideband microstrip patch antenna designed to cover 1.8 GHz – 2.7 GHz frequency range. A prototype having dimensions 400 x 400

mm^2 (approximately 3λ x 3λ) has been fabricated and tested. Four elementary sources constituting a 2 x 2 wideband patch array are used as primary source; the inter-element spacing being 200 mm.

Direct near field and far field performed measurements are shown in figure 4. When the capacitance of the metasurface reflector is changed by varying bias voltage of the varactor diodes, the frequency of maximum gain is tuned. As shown by the intensity maps of scanned far field, the emission frequency represented by the red spot varies from 1.9 GHz to 2.31 GHz from 0 V to 24 V, demonstrating clearly the frequency reconfigurability of the FP cavity. The radiation patterns in E- and H-planes at respectively 1.9 GHz, 2.02 GHz, 2.16 GHz and 2.31 GHz corresponding to maximum gain frequency for 0 V, 5 V, 12 V and 24 V are presented. The tuning range of maximum gain frequency results in an effective operation bandwidth close to 20%. A directivity of approximately 18 dBi is obtained experimentally.

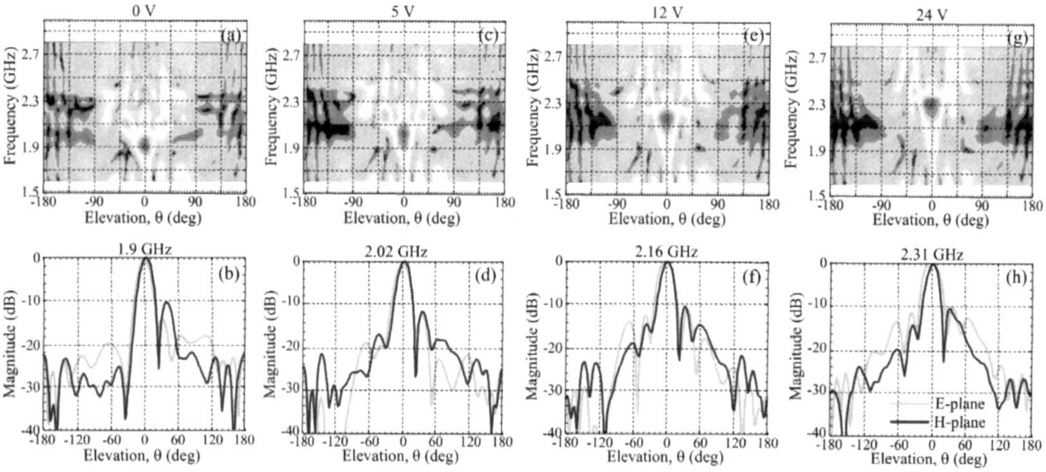

Figure 4. Far field intensity maps versus frequency and elevation angle θ in E-plane and measured radiation patterns in E- and H-planes at maximum gain frequency for different bias voltage applied: (a)-(b) 0 V – 1.9 GHz, (c)-(d) 5 V – 2.02 GHz, (e)-(f) 12 V – 2.16 GHz, (g)-(h) 24 V – 2.31 GHz.

5. Conclusion

To summarize, we have shown different characteristics that can be obtained from a Fabry-Pérot leaky wave antenna, namely, low profile, high directivity, beam steering and frequency agility, by the judicious configuration of the metamaterial-based PRS reflector.

References

[1] Trentini G V 1956 *IRE Trans. Antennas Propag.* **AP-4** 666
[2] Feresidis A P, Goussetis G, Wang S and Vardaxoglou J C 2005 *IEEE Trans. Antennas Propag.* **53** 209
[3] Zhou L, Li H, Qin Y, Wei Z and Chan C T 2005 *Appl. Phys. Lett.* **86** 101101
[4] Ourir A, de Lustrac A and Lourtioz J M 2006 *Appl. Phys. Lett.* **88** 084103
[5] Ourir A, de Lustrac A and Lourtioz J M 2006 *Microwave Opt. Technol. Lett.* **48** 2573
[6] Yahiaoui R, Burokur S N and de Lustrac A 2009 *Electron. Lett.* **45** 814
[7] Burokur S N, Yahiaoui R and de Lustrac A 2009 *Microwave Opt. Technol. Lett.* **51** 1883
[8] Ourir A, Burokur S N and de Lustrac A 2007 *Electron. Lett.* **43** 493
[9] Ourir A, Burokur S N, Yahiaoui R and de Lustrac A 2009 *C. R. Physique* **10** 414
[10] Ourir A, Burokur S N and de Lustrac A 2007 *Electron. Lett.* **43** 698
[11] Burokur S N, Daniel J P, Ratajczak P and de Lustrac A 2010 *Appl. Phys. Lett.* **97** 064101
[12] Burokur S N, Daniel J P, Ratajczak P and de Lustrac A 2011 *Microwave Opt. Technol. Lett.* **53** 2291

IOP Conf. Series: Materials Science and Engineering **44** (2013) 012014 doi:10.1088/1757-899X/44/1/012014

Electromagnetic waves reflectance from $TbMnO_3$ with sinusoidal antiferromagnetic structure

Igor V Bychkov [1], Dmitry A Kuzmin [1], Sergei J Lamekhov [1] and Vladimir G Shavrov [2]

[1] Chelyabinsk State University, 454001, Chelyabinsk, Br. Kashirinyh Street, 129, Russia

[2] The Institute of Radioengineering and Electronics of RAS, 125009, Moscow, Mokhovaya Street, 11-7, Russia

E-mail: bychkov@csu.ru

Abstract. The spectrum of coupled spin, acoustic, dipole and electromagnetic waves for $TbMnO_3$ with a sinusoidal antiferromagnetic structure is considered. The possibility of resonant interaction of spin, dipole, acoustic and electromagnetic waves is shown. The electromagnetic waves reflectance from the semi-infinite layer of $TbMnO_3$ with a sinusoidal antiferromagnetic structure is calculated. The permittivity of $TbMnO_3$ is investigated as well.

1. Introduction

Nowadays multiferroics (materials with magnetic and electrical ordering) attract researchers' attention with their special properties [1, 2]: existence of electromagnons and possibility to control them by an external electrical field. Many multiferroics have a modulated magnetic structure. It contributes to some features in the spectrum of excitations [3, 4]: band structure is observed, the nonreciprocity effect is manifested. Orthorhombic multiferroic $TbMnO_3$ has a sinusoidal antiferromagnetic structure at temperatures 28K < T < 41K. Coupled spin, dipole and electromagnetic waves in $TbMnO_3$ with sinusoidal antiferromagnetic structure have been investigated earlier [5], but influence of interaction of these subsystems with the acoustic one has not been studied. In the present work the spectrum of the coupled spin, dipole, acoustic and electromagnetic waves in $TbMnO_3$ with sinusoidal antiferromagnetic structure is investigated. Studies of spectrum of the coupled excitations in the modulated magnetic structures are made in case $L >> a$, where $L = 2\pi/k$ – the period of modulation, a – the lattice constant, k – the wave number of the modulated structure.

2. The spectrum of coupled waves

The method of Lagrange has been used for solving this problem. The Lagrangian's expression is $L=E-F$, where E – the kinetic energy, F – the Ginzburg-Landau functional. In this case:

$$E = \frac{1}{V}\int \frac{1}{2}\left(\mu\dot{\mathbf{A}}^2 + \lambda\dot{\mathbf{P}}^2 + \rho\dot{\mathbf{u}}^2\right)d\mathbf{r};$$
$$F = \frac{1}{V}\int \{\frac{a}{2}\mathbf{A}^2 + \frac{w}{2}A_z^{\ 2} + \frac{u}{4}\mathbf{A}^4 + \frac{\gamma}{2}\left(\partial_y\mathbf{A}\right)^2 + \frac{\alpha}{2}\left(\partial_y^2\mathbf{A}\right)^2 + \frac{b}{2}\mathbf{P}^2 - \mathbf{PE} + \tag{1}$$
$$+\nu P_z\left(A_z\partial_y A_y - A_y\partial_y A_z\right) + c_{ijlm}u_{ij}u_{lm} + b_{ijlm}A_iA_ju_{lm} + d_{ijlm}P_iP_ju_{lm} + f_{ijlm}u_{lm}P_i\partial_y A_j\}d\mathbf{r}.$$

Here $\mathbf{M}=\mathbf{M}_1+\mathbf{M}_2+\mathbf{M}_3+\mathbf{M}_4$ is the magnetization of the crystal; $\mathbf{A}=\mathbf{M}_1-\mathbf{M}_2-\mathbf{M}_3+\mathbf{M}_4$ is the vector of antiferromagnetism; $\mathbf{P}$ is the vector of polarization; $u_{ij}=\left(\partial u_i/\partial x_j+\partial u_j/\partial x_i\right)/2$ is the tensor of deformations; $\mathbf{u}$ is the displacement vector; w is the constant of anisotropy; $a,u,\beta,\lambda_1,\lambda_2$ are the constants of homogeneous exchange; α,γ are the constants of heterogeneous exchange; b,ν are the constants of electric and magnetoelectric interactions; $c_{ijlk},b_{ijlk},d_{ijlk},f_{ijlm}$ are the tensors of elasticity, magnetostriction, electrostriction and and electromagnitostriction; $\lambda=mv_c/z^2$, where z and m are the charge and the reduced mass of the elementary cell with the volume v_c; $\mu=\chi_\perp/8g^2M_0{}^2$, where $\chi_\perp$ is the static transversal magnetic susceptibility, g is the gyromagnetic ratio, M_0 is the magnetization of sub-lattices. Existance of term $f_{ijlm}u_{lm}P_i\partial_y A_j$ in (1) is due to the crystal symmetry [6].

The ground state of $TbMnO_3$ with sinusoidal antiferromagnetic structure is described by vectors of polarization $\mathbf{P_0}$ and antiferromagnetism $\mathbf{A_0}$ with the components: $P_{0x}=P_{0y}=P_{0z}=0;\ A_{0x}=A_{0z}=0;\ A_{y0}=A_1\cos ky$. From the minimum of Ginzburg-Landau functional with these expressions we obtain that $k^2=-\gamma/2\alpha;\ A_1{}^2=-4L_1/(3u-24\delta_{B2});\ L_1=a-\gamma^2/4\alpha;$ $\mathbf{B_2}=(B_{21},B_{22},B_{23});\ \delta_B=\Delta^{-1}\left(\Delta_{B21}+\Delta_{B22}+\Delta_{B23}\right);\ \Delta=\det\left|c_{ij}\right|$, Δ_{B2i} can be gotten from Δ by changing column with number i to vector $\mathbf{B_2}$.

For getting the spectrum of coupled waves one has to take into account the system of Maxwell's and Lagrange equations for **A**, **P** and **u**.

Let us now investigate the incidence of electromagnetic wave with $\mathbf{e}\|Z$, so $e_z=e_0\exp[i(qy-\omega t)]$. Solving the system of equations by the method of low oscillations, linearizing it and using the form of harmonic series for variables, in approach of first harmonics we will get a dispersion equation. With the approximation $q \ll k$ it has an analytical form:

$$\begin{aligned}&\omega_{fpa}^4\left(\omega^2-\omega_l^2\right)\left\{\left(\omega^2-\tilde{\omega}_0^2\right)\left(\omega^2-\omega_t^2\right)-\omega_{mel}^2\omega_t^2\right\}+\left(\omega^2-\omega_{0y}^2\right)\left\{\left[\left(\omega^2-\tilde{\omega}_0^2\right)\left(\omega^2-\omega_t^2\right)-\omega_{mel}^2\omega_t^2\right]\times\\&\times\left[\left(\omega^2-\omega_p^2+\frac{\omega_E^2}{n^2-1}\right)\left(\omega^2-\omega_l^2\right)+\omega_{fp}^2\omega_l^2\right]-\omega_{me}^4\left(\omega^2-\omega_t^2\right)\left(\omega^2-\omega_l^2\right)\right\}=0\end{aligned}\tag{2}$$

Here, notations from appendix A are considered.

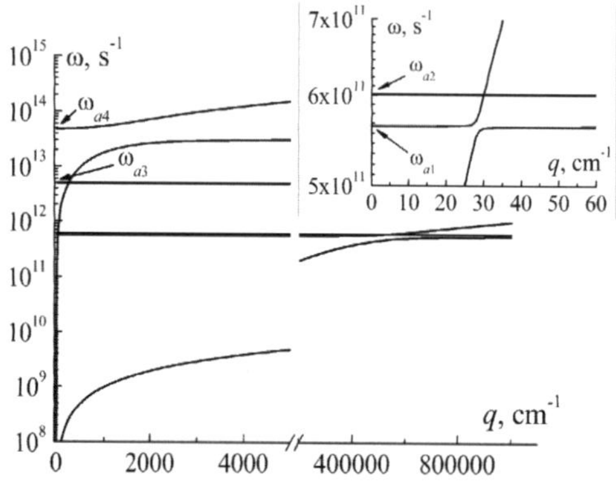

Figure 1. The spectrum of coupled waves.

Using typical values of constants for $TbMnO_3$: g = $2\cdot10^7$ s^{-1} erg^{-1}, M_0 ~ 10^3 Oe, α ~ 10^{-28} cm^4, γ ~ -10^{-14} cm^2, q ~ 10^7 1/cm, a ~ 1, $s_t=3\cdot10^5$ cm/s, $s_l=5\cdot10^5$ cm/s, ρ = 8 g/cм^3, μ ~ 10^{-23} s^2, λ ~ 10^{-26} s^2

from equation (2) we will obtain the spectrum of coupled spin, elastic and electromagnetic excitations in $TbMnO_3$ with sinusoidal antiferromagnetic structure (Fig. 1).

Figure 1 shows that the spectrum has a band structure. Band gaps are observed for both electromagnetic and elastic excitations. There are four activation frequencies in the spectrum. $\omega_{a3} = s_l k$ and $\omega_{a2} = \mu^{-1/2}\left(w + a + 3uA_1^2/2 + 4A_1^2\delta_{B2}\right)^{1/2}$ belong to longitudinal acoustic and spin waves, respectively; $\omega_{a1}^2 \approx \dfrac{\omega_0^2\left(\Omega_p^2 + \omega_{fp}^2\right) + \omega_{me}^4}{\Omega_p^2 + \omega_{fp}^2 + \tilde{\omega}_0^2}$ and $\omega_{a4}^2 \approx \Omega_p^2 + \left(\omega_{me}^4 - \omega_l^2\omega_{fp}^2\right)^{1/2}$ (read appendix A for notations) are caused by interaction of electromagnetic waves with spin and dipole subsystems, respectively. There is a "repulsion" of spin and elastic modes of oscillations and as the result the magnetoacoustic gap exists. Interaction of subsystems has a resonant type and appears mostly near points of crossing of non-interacting oscillations modes, and as a result band gaps are observed. Bands widths depend on intensity of interaction.

3. Electromagnetic waves reflectance from $TbMnO_3$ with sinusoidal antiferromagnetic structure

Let us now investigate the reflection of linear polarized electromagnetic waves with $\boldsymbol{e}\|Z$ and $e_z = e_0\exp[i(qy-\omega t)]$ from a semi-infinite layer $TbMnO_3$ with sinusoidal antiferromagnetic structure. Consider normal incidence of electromagnetic wave. The solution of system of boundary conditions with dispersion equation (2) gives the reflectance $R = \left|e_R/e_0\right|^2$. Fig. 2 shows the frequency dependence of this reflectance.

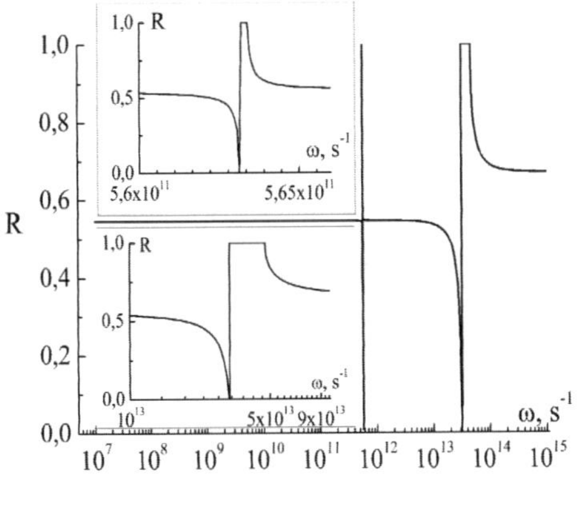

Figure 2. Frequency dependence of reflection coefficient of electromagnetic waves.

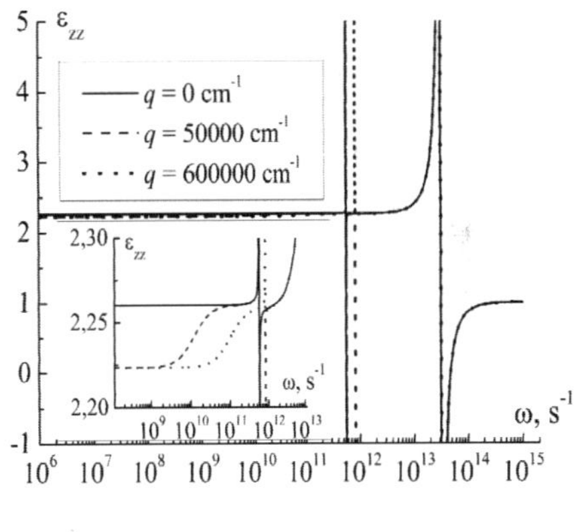

Figure 3. Frequency dependencies of ε_{zz} for different wave numbers.

4. Permittivity

Whenever electromagnetostriction interaction is very weak, we will put $f_1 = 0$ in the equation of dispersion (2). Expression for $n^2 = \varepsilon_{zz}$ has the form:

$$\varepsilon_{zz} = n^2 = \frac{\left(\omega^2 - \Omega_1^2\right)\left(\omega^2 - \Omega_2^2\right)\left(\omega^2 - \Omega_3^2\right)\left(\omega^2 - \Omega_4^2\right)}{\left(\omega^2 - \omega_1^2\right)\left(\omega^2 - \omega_2^2\right)\left(\omega^2 - \omega_3^2\right)\left(\omega^2 - \omega_4^2\right)}, \tag{3}$$

where values of Ω_i are obtained from the system of equations:

$$
\begin{aligned}
&\Omega_1^2+\Omega_2^2+\Omega_3^2+\Omega_4^2=\tilde{\omega}_0^2+\Omega_p^2+\omega_l^2+\omega_t^2;\\
&\Omega_1^2\left(\Omega_2^2+\Omega_3^2+\Omega_4^2\right)+\Omega_2^2\left(\Omega_3^2+\Omega_4^2\right)+\Omega_3^2\Omega_4^2=\left(\tilde{\omega}_0^2+\omega_t^2\right)\left(\Omega_p^2+\omega_l^2\right)-\omega_{me}^4;\\
&\Omega_1^2\Omega_2^2\left(\Omega_3^2+\Omega_4^2\right)+\Omega_3^2\Omega_4^2\left(\Omega_1^2+\Omega_2^2\right)=\omega_{me}^4\left(\omega_l^2+\omega_t^2\right)-\omega_l^2\Omega_p^2\left(\tilde{\omega}_0^2+\omega_t^2\right)-\omega_t^2\left(\Omega_p^2+\omega_l^2\right)\left(\tilde{\omega}_0^2-\omega_{mel}^2\right);\\
&\Omega_1^2\Omega_2^2\Omega_3^2\Omega_4^2=\omega_l^2\omega_t^2\left\{\Omega_p^2\left(\tilde{\omega}_0^2-\omega_{mel}^2\right)-\omega_{me}^4\right\}.
\end{aligned}
\tag{4}
$$

ω_i is obtained like Ω_i if we puts $\omega_E^2=0$ in (4).

Expression (4) depends on both frequency and wave number, so time and space dispersions in $TbMnO_3$ coexist.

Fig. 3 shows the frequency dependence of permittivity for different wave numbers.

5. Summary

In the present work, influence of subsystem interaction on spectrum and dynamic characteristics of $TbMnO_3$ with sinusoidal antiferromagnetic structure have been considered. The analysis has shown that the spectrum of coupled waves has a band structure. Interaction of subsystems has a resonance type and takes maximal values close to points of crossing modes of non-interacting oscillations. This is the reason of band gaps. Permittivity depends on frequency and wave number, i.e. time and space dispersion coexist in $TbMnO_3$. There are peaks in ε_{zz} frequency dependence. Calculating the frequency dependence for reflection of electromagnetic waves from surface of semi-infinite $TbMnO_3$ with sinusoidal structure shows existence of opacity windows in some intervals of frequencies.

Acknowledgments

The work was supported by grant of RFBR № 11-02-12255.

Appendix A. List of notations

The following notations are used: $\tilde{\tilde{u}}=u+8\delta_{B3}$, $\omega_p{}^2=\lambda^{-1}\left(b-4A_1{}^2\delta_{D3}\right)$, $\omega_0^{\pm\,2}=\mu^{-1}\left(w+a+\tilde{\tilde{u}}A_1{}^2/2+\gamma(q\pm k)^2+\alpha(q\pm k)^4\right)$, $\omega_{l\pm}^2=s_l{}^2(q\pm k)^2$, $\omega_t^2=s_t{}^2q^2$, $s_l{}^2=c_{22}/\rho$, $s_t{}^2=2c_{44}/\rho$, $\omega_{0y}{}^2=\mu^{-1}\left(w+a+3uA_1{}^2/2+4A_1{}^2\delta_{B2}+\gamma q^2+\alpha q^4\right)$, $\omega_{me}^4=2\nu^2k^2A_1^2/\lambda\mu$, $\omega_{fp}^2=f_1{}^2A_1k^2/\mu c_{22}$, $\omega_E^2=4\pi/\lambda$, $\omega_l=s_lk$, $\omega_0^2=\mu^{-1}\left(w+a+\tilde{\tilde{u}}A_1{}^2/2+\gamma k^2+\alpha k^4\right)$, $\tilde{\omega}_0^2=\omega_0^2-\tilde{\tilde{u}}A_1{}^2/4$, $\omega_{fpa}^4=f_1{}^2A_1^4q^2\delta_{B3}/2\mu\lambda$, $\omega_{mel}^2=A_1^2B_{44}^2/4\mu c_{44}$, $n^2=qc/\omega$, n – the refractive index, $\omega_0^2=\mu^{-1}\left(w+a+\tilde{\tilde{u}}A_1{}^2/2+\gamma k^2+\alpha k^4\right)$, $\Omega_p^2=\omega_p^2+4\pi/\lambda$.

References

[1] Markku P V Stenberg and R. de Sousa 2012 *Phys. Rev. B* **85** 104412
[2] Rovillain P, Cazayous M, Gallais Y, Measson M-A, Sacuto A, Sakata H, and Mochizuki M 2011 *Phys. Rev. Lett.* **107** 027202
[3] Bychkov I V, Kuzmin D A, Shavrov V G 2013 *J Magn. Magn. Mater.* **329** 142
[4] Binbin Wu and G J Diebold 2012 *Appl. Phys. Lett.* **100** 164102
[5] Chupis I E 2007 *Low Temp. Phys.* **33** 715
[6] Ribeiro J L and Vieira L G 2010 *Phys. Rev. B* **82** 064410

Accurate Measurement of RF Exposure from Emerging Wireless Communication Systems

Thierry Letertre[1], Vikass Monebhurrun[2] and Zeno Toffano[1]
[1]Department of Telecommunications, SUPELEC, 3 rue Joliot-Curie, 91192 Gif sur Yvette Cedex, France
[2]Department of Electromagnetics, DRE-L2S, SUPELEC, 3 rue Joliot-Curie, 91192 Gif sur Yvette Cedex, France

E-mail: thierry.letertre@supelec.fr, vikass.monebhurrun@supelec.fr and zeno.toffano@supelec.fr

Abstract. Isotropic broadband probes or spectrum analyzers (SAs) may be used for the measurement of rapidly varying electromagnetic fields generated by emerging wireless communication systems. In this paper this problematic is investigated by comparing the responses measured by two different isotropic broadband probes typically used to perform electric field (E-field) evaluations. The broadband probes are submitted to signals with variable duty cycles (DC) and crest factors (CF) either with or without Orthogonal Frequency Division Multiplexing (OFDM) modulation but with the same root-mean-square (RMS) power. The two probes do not provide accurate enough results for deterministic signals such as Worldwide Interoperability for Microwave Access (WIMAX) or Long Term Evolution (LTE) as well as for non-deterministic signals such as Wireless Fidelity (WiFi). The legacy measurement protocols should be adapted to cope for the emerging wireless communication technologies based on the OFDM modulation scheme. This is not easily achieved except when the statistics of the RF emission are well known. In this case the measurement errors are shown to be systematic and a correction factor or calibration can be applied to obtain a good approximation of the total RMS power.

1. Introduction

With the proliferation of wireless communication systems, there is an ever-increasing public concern regarding human exposure to RF fields. Furthermore, in order to avoid potential electromagnetic interference among wireless communication devices, there is also a need to control the electromagnetic emissions. Herein the in-situ measurement of the electric fields (E-fields) emitted by current or future wireless communication systems is investigated. An isotropic broadband probe or a spectrum analyzer (SA) may be used for the in-situ RF exposure evaluation [1]. The legacy measurement protocols are well adapted for the E-field evaluations of regular or predictable systems such as the 2g and the 3g wireless communication systems that occupy a frequency bandwidth of up to 5MHz with a relatively small variation of the signal envelope. Indeed such signal characteristics are compatible with currently used isotropic broadband E-field measurement systems. However the measurement protocols prove to be inadequate for signals with high CF that are emitted by emerging wireless communication systems based on the OFDM modulation scheme such as WiFi, WIMAX and LTE systems. Several recent works have addressed the issue of the E-field measurements of complex digital signals using measurement instruments that are typically calibrated using continuous wave

(CW) signals [2]. The effect of the digital modulations on the measurements of the specific absorption rate (SAR) is discussed in [3] whereas the in-situ measurements using personal exposimeters are discussed in [4]. The influence of the calibration parameters on the in-situ measurements using either selective or non-selective equipment is discussed in [5]. The influence of the digital modulation schemes (e.g. Phase Shift Keying (PSK), Quadrature Amplitude Modulation (QAM), and OFDM) on the E-field measurements is further discussed in [6].
In emerging wireless communication systems, the CF, also referred to as the peak to average power ratio (PAPR), is a key parameter that must be taken into account because of its non-negligible impact on the E-field evaluations [7-8]. WIMAX and LTE signals are rather difficult to measure accurately because of the rapid variations of their temporal shape. Typically two types of measurement errors have been identified. The first error which is a deterministic one is due to the "burst" and/or due to the duty cycle (DC) of the signal thereby producing a strong power variation over a relatively short time (within the frame time) relative to its average value (typically a few ms). This affects the average power measurements since most SAs and isotropic broadband probes cannot take into account instantaneously the emitted power because of the frame duration which is shorter than the rise time of the detector. The second error which is non-deterministic is inherent to the use of the OFDM modulation scheme in association with the Orthogonal Frequency Division Multiple Access (OFDMA) technique. The superposition of several modulated subcarriers over a wide bandwidth results in relatively high and variable PAPR values (ranging from 10dB to 20dB).
To account for these two errors, a calibration of the measurement system using complex test signals becomes a prerequisite.

2. Test bench and protocol descriptions

The experimental setup is shown in Figure 1. Two isotropic broadband probes, labelled A and B, are submitted to four different signal configurations emitted with the same carrier frequency:

a) Pure sinusoidal or CW signal, with or without pulsed mode,
b) Single-carrier modulated signal,
c) Non-modulated multi-tone signal with variable number of CW sub-carriers,
d) WIMAX or LTE standard compliant signal where the number of OFDM symbols of the User Data sub-frame is chosen to achieve a given DC target value.

The signals are synthetized using a vector signal generator (Agilent ESG 4438C with the "Signal Studio" software ©Agilent Technologies). Prior to the radiated mode measurements using the isotropic broadband probes, and in aim to minimize the errors, a SA is first used in the conducted mode to verify that the output RMS power of the signal generator is unchanged whatever the waveform chosen. An attenuator is used to reduce the power measured by the SA and compensation is also allowed for the additional cable loss. The embedded complementary cumulative distribution function (CCDF) command of the SA is applied to measure the PAPR or CF values of the emitted waveform. The working frequency ($f = 2.45$GHz) is arbitrarily selected to ensure compatibility with the measurement equipment. The distance d ($d > \lambda \approx 12.3$cm) and the RMS output power ($P_{RMS} = +20$dBm) of the signal generator are fixed to ensure a sufficient level of the E-field for the two isotropic broadband probes without reaching the saturation level for the radiated mode measurements.

Figure 1. Schematic of the Test Bench wherein a SA and two commercially available broadband probes are considered: a selective and isotropic personal exposure meter incorporating 12 frequency bands (providing instantaneous and mean E-field measurements) labelled Probe A and a broadband field meter with isotropic E-field probe (providing true RMS E-field measurements) labelled Probe B.

3. Measurement results

Figure 2 shows the responses of Probes A and B for the cases of pulsed and WIMAX signals with variable DCs. The results show that the two isotropic broadband probes under test respond correctly to pulsed and CW (DC=1) signals. The E-field values measured by Probe A correspond to the peak value and they are independent of the temporal shape of the signal for pulse widths greater than 4ms. On the other hand the E-field values measured by Probe B correspond to the RMS value and they are linearly dependent on the temporal shape or the DC value of the signal.

Figure 2. Comparison of the E-fields measured using Probes A and B for the cases of pulsed and WIMAX signals with variable DCs.

When the two probes are submitted to WIMAX stimuli with the same DC as the previously considered pulsed signals (10% < DC < 50%), the measured E-fields are underestimated. The systematic errors are found to be linearly dependent on the square root of the DC value. They can therefore be corrected for the case of deterministic signals with known DC.

Figure 3(a) and (b) show the CCDF and Channel Power measurements of the two waveforms using the SA. The responses obtained by the SA using the default operating mode (e.g. Trace in “ClearWrite” mode) are highly perturbed by the temporal variations of the signal and by the DC value. Strobe effects occur between the applied frequency sweep and the “burst” or frames period. Furthermore the measured power is underestimated when the signal bandwidth (from 1 to 20MHz for WIMAX or LTE) is greater than the resolution bandwidth (RBW) of the SA. Some SAs offer alternative measurement capabilities such as “ChPwr” (Channel Power) or CCDF calculus, which can circumvent the small value of the RBW as follows:

a) ChPwr mode: Measurement of the total average power in a specific or chosen channel bandwidth. The measurement is reliable for "broadband", “continuous" or constant DC signals (e.g. WCDMA).

b) CCDF mode: It provides information on how often the measured power is above a given value (typically the average value). A good estimation of the average value of the signal and the PAPR is obtained.

From Figures 3(a) and (b) it is observed that the errors are constant and can be easily adjusted by applying a correction factor.

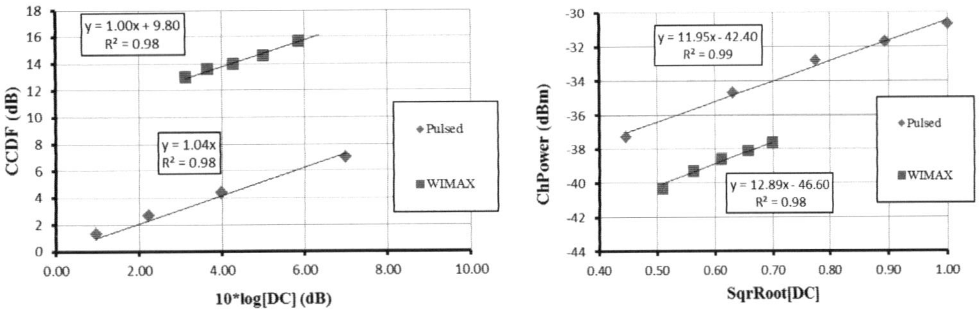

Figure 3. (a) CCDF (left) and (b) ChPwr (right) measurements of the pulsed and WIMAX waveforms using the SA.

4. Conclusion

The E-field measurements using isotropic broadband probes show that such diode-detector based probes are not suitable for signals with relatively high power and time variations. A high value of the CF reduces the sensitivity and increases the uncertainty of the measurement. Furthermore the rise time of these detectors is not fast enough to cope with the rapid variations of modern wireless communication signals. The OFDM modulation scheme is not solely responsible for the measurement errors; the wide bandwidth of the useful frequency band is another cause. In order to accurately measure such signals, two approaches may be adopted:

1) Improve the detecting capabilities of isotropic broadband probes in order to render them compatible with the rapid temporal variations of the signals.
2) Apply a first correction factor in order to compensate for the CW-based calibration for the case of deterministic signals (such as WIMAX and LTE) and a second correction factor which takes into account the traffic statistics for non-deterministic signals (such as WiFi).

To compensate for the under-estimation of WIMAX signals due to the response time of the probes, the measured E-field values should be herein multiplied by correction factors $\alpha = 2.78$ and $\beta = 2.08$ for Probes A and B, respectively. Furthermore when using the SA for the in-situ measurements, the more reliable method consists in using the channel power mode, with a very long sweep time (>> frame duration) over the signal bandwidth, to obtain the RMS power. The peak power value may then be estimated by taking in account the PAPR or the CF. However an overestimated value will obtained for the case of QAM modulation. An alternative solution is to use the CCDF measurement mode of the SA when it is available.

5. References

[1] CENELEC 2008 Basic standard for the in-situ measurement of electromagnetic field strength related to human exposure in the vicinity of base stations *EN-50492*

[2] IEEE 2005 Standard for Calibration of Electromagnetic Field Sensors and Probes, Excluding Antennas, from 9kHz to 40GHz IEEE Std 1309-2005

[3] Monebhurrun V 2010 Effect of time-averaging of pulsed radio-frequency signals on Specific Absorption Rate measurements *IEEE Trans. On Electromagnetic Compatibility* vol 52 no 1 pp 49-55

[4] Adamson D Bownds D, Fernandez A and Goodal E. 2010 The response of Electric Field probes to realistic RF environments IEEE MTT-S (Anaheim, CA, USA, May 2010)

[5] Sarolic A Roje V and Modlic B 2006 Measurement of electric field probe error for pulsed signals *Proc. 2006 IEEE Intl Symposium on Electromagnetic Compatibility* vol 2 pp 244-248 (Portland, OR, USA, August 2006)

[6] Letertre T, Monebhurrun V and Toffano Z 2011 Electromagnetic Field Measurements of Wimax Systems Using Isotropic Broadband Probes IMS 2011 (Baltimore, USA, June 2011)

[7] Wunder G and Paterson K 2004 Crest-factor analysis of carrier interferometry MC-CDMA and OFDM systems *Proc. 2004 IEEE International Symposium on Information Theory* (Chicago, ILL, USA, July 2004)

[8] Joseph W, Verloock L, Goeminne F, Vermeeren G and Martens L 2012 Assessment of RF exposure from emerging wireless communication technologies in different environments *Health Physics 2012* vol 102 no 2 pp 161-172

Output power distributions of mobile radio base stations based on network measurements

D Colombi[1], B Thors[1], T Persson[1], N Wirén[2], L-E Larsson[2] and C Törnevik[1]

[1]Ericsson Research, SE-164 80 Stockholm, SWEDEN
[2]TeliaSonera, SE-123 43 Farsta, SWEDEN

E-mail: davide.colombi@ericsson.com

Abstract. In this work output power distributions of mobile radio base stations have been analyzed for 2G and 3G telecommunication systems. The approach is based on measurements in selected networks using performance surveillance tools part of the network Operational Support System (OSS). For the 3G network considered, direct measurements of output power levels were possible, while for the 2G networks, output power levels were estimated from measurements of traffic volumes. Both voice and data services were included in the investigation. Measurements were conducted for large geographical areas, to ensure good overall statistics, as well as for smaller areas to investigate the impact of different environments. For high traffic hours, the 90th percentile of the averaged output power was found to be below 65% and 45% of the available output power for the 2G and 3G systems, respectively.

1. Introduction

Human exposure to radio frequency (RF) electromagnetic fields (EMF) from base stations is directly proportional to the base station output power. In determining EMF limit compliance, the theoretical maximum output power is normally used. In reality, however, the transmitted power can be much lower than the maximum levels depending on factors such as power regulation, discontinuous transmission (DTX), traffic variations and radio access technology [1]. The objective of this study was to gather statistics on downlink output power in mobile networks to determine typical power levels and realistic distributions.

2. Method

In mobile communication networks it is possible to monitor many performance indicators to ensure that the networks meet expected quality requirements [2]. The performance surveillance is part of the Operational Support Systems (OSS) used by the network operators. Depending on access technology there are counters that either directly or indirectly can be employed to determine the downlink output power. Using these types of counters it was possible to simultaneously gather statistics for large sets of radio base stations (RBS) in the TeliaSonera 2G and 3G networks in Sweden. Effects of traffic variation and different environments such as urban and rural areas were analyzed for different radio access technologies

The data monitored were sampled by the system with a relatively narrow time interval, about 100 ms for 3G and 10 s for 2G, and averaged over 15 minutes before being transmitted to the OSS. Thus, the analyzed figures correspond to RBS output power levels averaged over 15 minutes.

2.1. 2G network data

Traffic volume data were collected for both GSM 900 and GSM 1800 during 24 hours of a weekday. The dataset included a total of 12400 sectors (or GSM cells), covering 35% of the land area and potentially 70% of the population. Traffic volume data were also gathered during a full week for two different base station controllers corresponding to rural and urban environments, respectively.

Both voice and data (GPRS+EDGE) traffic were included in the analysis. The traffic volume data were translated to output power by calculating an average number of TRXs used during each 15 minute period. In this model, effects such as discontinuous transmission (DTX), frequency hopping, and number of traffic channels used for GPRS and EDGE were considered. The impact of DTX was taken into account by multiplying the obtained power levels related to voice traffic with a factor of 0.6 [3].

The distribution of the transmitted power compared to the maximum available power from the RBS was determined as function of the number of installed transceivers (TRXs). In GSM, TRXs are responsible for the transmission and reception of the signal over a pair of 200 kHz channels. Increasing the number of TRXs will increase the cell capacity.

2.2. 3G network data

Output power data were collected in the WCDMA 2100 MHz band during 24 hours of a weekday. Data for a total of three Radio Network Controllers (RNCs) were included, corresponding to about 220 radio base stations, covering urban, suburban, and rural areas.

3. Results

In the subsections below, results obtained for the 2G and 3G networks are presented.

3.1. 2G

In 99% of the analyzed GSM cells, the number of installed TRXs was less than seven, with cells containing two TRXs constituting the clear majority. The empirical probability density function (PDF) of the output power for the 2G network, including all collected data over 24 hours, is shown in Figure 1a. Figure 1b shows the corresponding cumulative distribution (CDF).

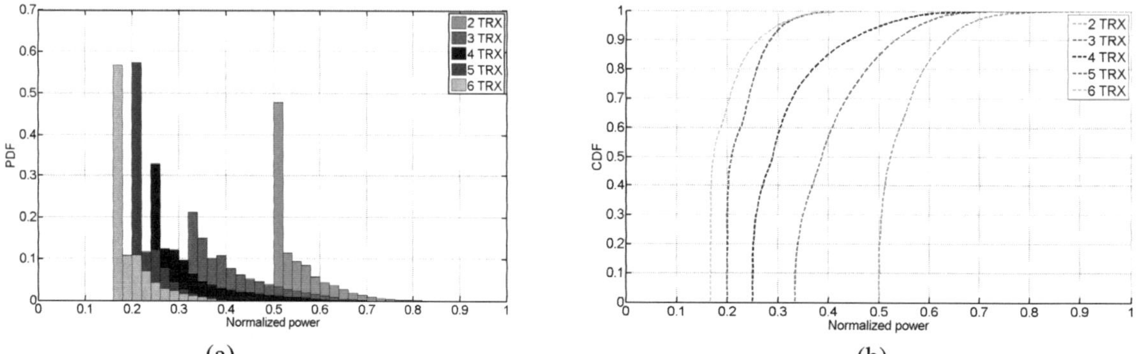

Figure 1: Normalized RBS output power for different number of installed TRXs in the considered 2G network. (a) Empirical Probability Density Function (PDF). (b) Empirical Cumulative Distribution Function (CDF).

The two first timeslots of the first TRX are normally used for signaling and this carrier is transmitted at constant power at all times. For those of the remaining six timeslots not used for traffic, dummy bursts are sent to ensure continuity in power levels [4]. As a consequence, at least one carrier is always transmitting at maximum power which, in combination with the employed normalization to maximum power levels, explains why the bars/curves in Figure 1 are shifted towards the left when more TRXs are added.

The CDFs in Figures 2a and 2b, show the differences in the output power distribution for low and high traffic hours (from 1 to 5 am and from 4 to 8 pm, respectively) and between urban and rural environments, respectively.

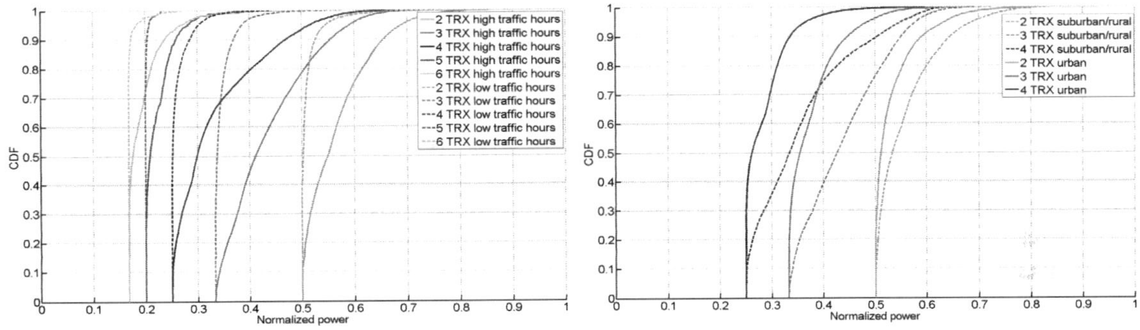

(a) (b)

Figure 2: CDFs of output power levels for different number of installed TRXs in the considered 2G network. (a) Variation between low and high traffic hours (all collected data). (b) Variation between different environments (one base station controller per environment).

3.2. 3G

The overall 3G output power statistics are given in Figure 3.

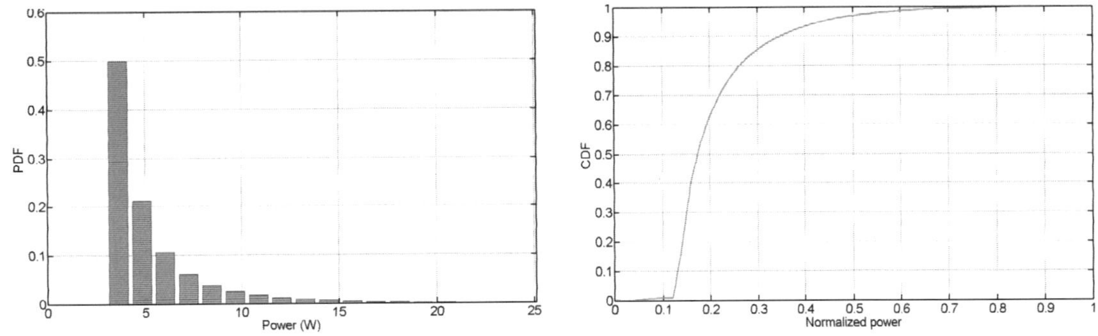

(a) (b)

Figure 3: 3G output power statistics for all sites investigated. (a) PDF of RBS output power per cell. (b) CDF of normalized output power with respect to the maximum value.

Figures 4a and 4b provide CDFs of the normalized output power for high versus low traffic hours (from 4 to 8 pm and from 1 to 5 am, respectively) and for different environments, respectively.

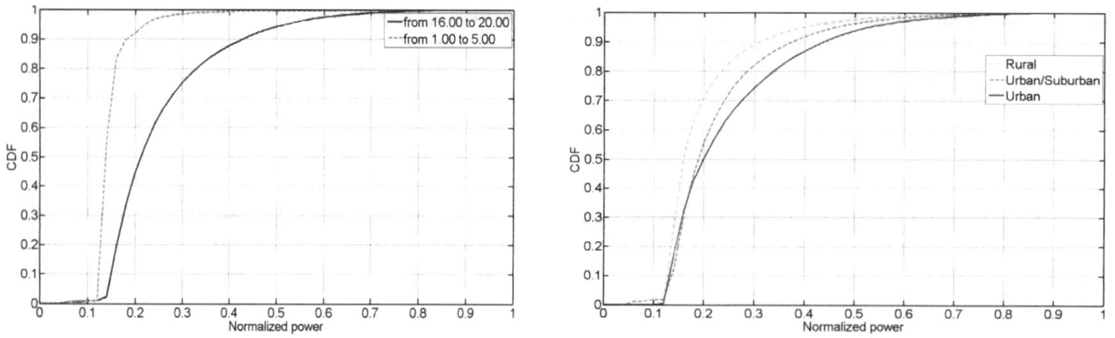

Figure 4: CDFs of output power levels for the considered 3G network. (a) Comparison between high and low traffic hours. (b) Comparison between different environments.

4. Discussion

4.1. 2G

For two TRxs and the dataset corresponding to the 24 hour measurements, the 90th percentile was found to be about 63% of the available output power. By restricting the measurements to high traffic hours, the corresponding figure increased to 65% as shown in Figure 2a. Similarly, in environments characterized primarily by suburban/rural areas (considering data for 24 hours), the 90th percentile output power was found to be 65%, see Figure 2b.

A characteristic of the investigated GSM network is that no power control technique is used. Therefore, the output power variation is mainly dependent on the traffic variation. In many other networks, however, power control algorithms are implemented to reduce interference and to save energy. With power control implemented, even lower RBS power levels would have been obtained.

4.2. 3G

The average output power per cell was 6 W, *cf.* Figure 3a. For 90% of the analyzed cells the maximum available output power was about 25 W. Including all collected power samples, the 90th percentile was found to be 35% of the available RBS output power. The corresponding figure considering high traffic hours only was 42%. By restricting the measurements to urban environments (all 24 hours included), the 90th percentile power level was 43% of the maximum output power.

5. Conclusion

In this work, large amount of network data were collected and the downlink output power distributions for 2G and 3G mobile communication networks were analyzed. Due to effects of power regulation, DTX, and traffic variations, the effective RBS power was found to be far below the maximum available power. For high traffic hours, the 90th percentile of the averaged output power was found to be 65% and 45% of the available RBS power for 2G and 3G, respectively.

References

[1] Mahfouz Z, Gati A, Lautru D, Wong M-F, Wiart J and V Hanna 2012 Influence of traffic variations on exposure to wireless signals in realistic environments *Bioelectromagnetics* **33** 288-297

[2] Holm-Öste G and Norling M 2008 Ericsson's user service performance framework, *Ericsson Review* 43-46

[3] 3GPP 2001 Technical specification group services and system aspects; full rate speech; voice activity detector (VAD) for full rate speech traffic channels *3GPP TS 46.032* **4.0.0**

[4] 3GPP 2009 Technical specification group GSM/EDGE radio access network; radio subsystem link control *3GPPTS 45*.008 **4.18.0**

Hand Effect on Head Specific Absorption Rate (SAR) Exposed By Two Realistic Phone Models

J Keshvari[1] & M Kivento[2]

1 Nokia Corporation, Corporate Development Office, Keilahdentie 2-4, 00240 Espoo-Finland
2 Tampere University of Technology, Department of Biomedical engineering Tampere-Finland

E-mail: Jafar .keshvari@nokia.com

Abstract. There have been some reports about possible effect of the hand presence on the head SAR if hand phantom is included in the measurements of the head SAR compliance assessment procedure. The objective of this computational study was to examine the reported effect by using realistic head models and realistic CAD based phone models. A commercially available FDTD based EM solver was used to carry out the computational work. Based on the results of this study considering the SAR values without hand phantom as reference, following conclusions can be made: 1. In general presence of the hand lead to significantly less conservative SAR values in the head for large majority of cases 2. For lower band GSM frequencies the presence of the hand decreases the head SAR up to ~70%. 3. For the upper band GSM frequencies the presence of the hand decreases the head SAR up to ~55%. Based on the results of this study the present SAR compliance protocol where hand phantom is not included leads to more conservative head SAR results compared to the cases where hand is included.

1. Introduction

Compliance of the wireless devices is routinely assessed in laboratories by measuring head SAR in a Standard Anthropomorphic Model (SAM) using a computer-controlled robot system inside a human head phantom filled with tissue-equivalent liquid without including the hand phantom in the measurement. It was assumed that exclusion of the hand will lead to a more conservative assessment of SAR in the head. There have been some reports [1, 2, and 4] about possible effect on the head SAR if hand is included in the compliance measurement procedure in the IEC 62209 or IEEE 1528 standard [3, 6]. The IEC MT1-62209-1 committee executed two large projects in order to study the reported effect both experimentally and computationally. The objective of those projects by IEC MT62209-1 was to verify conservative nature of SAM phantom.The measurement study was lacking the use of real heads and the computational study did not use real mobile phones in their simulations. Because simulations with CAD based realistic phone models and MRI based head models are the closest exposure scenario representing real human head exposures we carried out this study to see how the inclusion of the hand will affect the head SAR in realistic exposure scenarios.

[1] Correspondence

2. Materials and methods
In this section we will describe the head models, the head models, the standardized head phantom (SAM) and the mobile phone CAD models.

2.1. Head models
Besides SAM phantom 9 anatomical head models were used. Most of the models are based on Virtual Family set which are developed by ITIS foundation and founded by Mobile manufacturer's forum (MMF). Children models include male Child 3yo, male Child 7yo, male child 6yo, and female child 8yo, female 11yo. Furthermore two adult male and two adult female models were included in the simulations. The CAD model of the SAM phantom is taken from ICES SCC34 website. The electric properties of the homogeneous tissue simulant material were set to the target values from IEEE 1528b (Draft version, May, 2010) as shown in Table 1:

Table 1: Target values from IEEE 1528b.

	P1528b Targets	
	ε_r	σ (S/m)
900 MHz	41.5	0.97
1800 MHz	40.0	1.4
2450 MHz	39.2	1.8
3500 MHz	38.0	2.91
5800 MHz	35.3	5.27

2.2. Hand models
The following hand models were used in this study: Two CTIA hand models for adult head models (Figure 1) and MRI based child hand model for child head models. The hand models grip provided representative practical way of holding each of the simulated transmitter models. For simulations with hand and head models the mobile phone was be positioned against the head the way as described in [5].

2.3. Mobile phone CAD models
Two Nokia mobile phone models (Figure 1) 6630 (GSM 900 / 1800, WCDMA 1950) and 8310 (GSM 900 / 1800) were used. The models are detailed CAD which are comprised of hundreds parts. The modelling phase of the mobile phones has been the most crucial phase of the computational work.

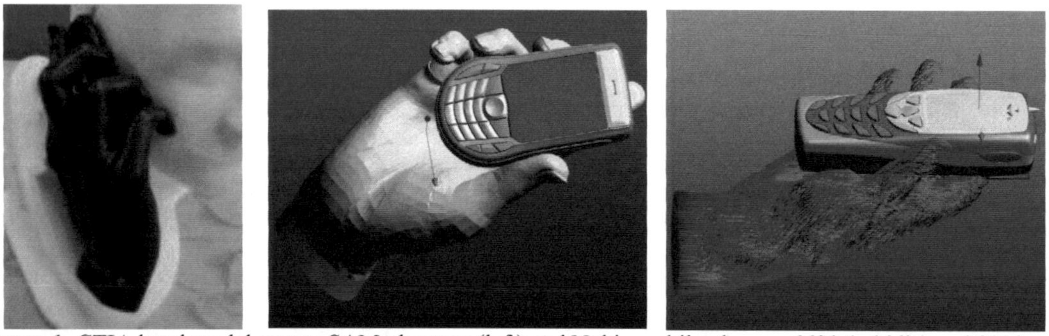

Figure 1. CTIA hand model next to SAM phantom (left) and Nokia mobile phones, 6630 (middle) and 8310 (right) with two different common hand grips.

3. Simulations, frequency bands and normalization of the results
Simulations were carried out using commercially available FDTD based software. The study is based on comparison of the SAR values computed in a standard SAR compliance test configurations with SAR values computed in the conditions when the same mobile phone models are held in the hand against the anatomically correct head models. The frequencies covered by this study are GSM frequencies 900 MHz, 1800 MHz, and

UMTS 1950 MHz. Spatial peak average of 1g and 10g SAR values were calculated for each configuration with SAM phantom and the results were normalized to 1 W antenna input power. Also spatial peak average 1g and 10g SAR values were calculated for head only tissues and head tissue with pinna for each configuration with anatomical head models which were normalized to 1W input power.

4. Results and conclusions

Because the amount of results data is large, only some examples will be provided in this paper. For the lower frequency, 900 MHz, the results for both phone models for two adult and two child head models in touch position are given in Figure 2. The values are expressed in percentile compared to reference value (SAM phantom without hand). The result for 1800 MHz is given in Figure 3. Head only tissue means that pinna tissues are not included in the SAR calculations.

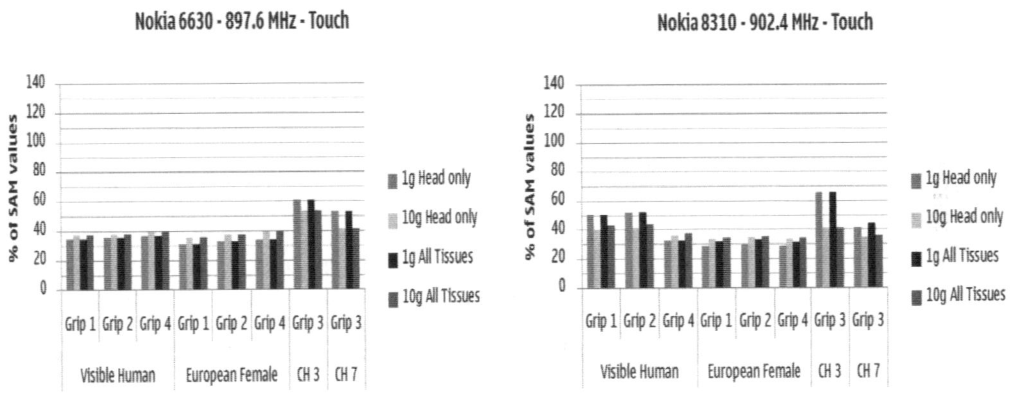

Figure 2. Relative SAR values for Nokia 6630 model (left) and Nokia 8310 model (right) at 900 MHz

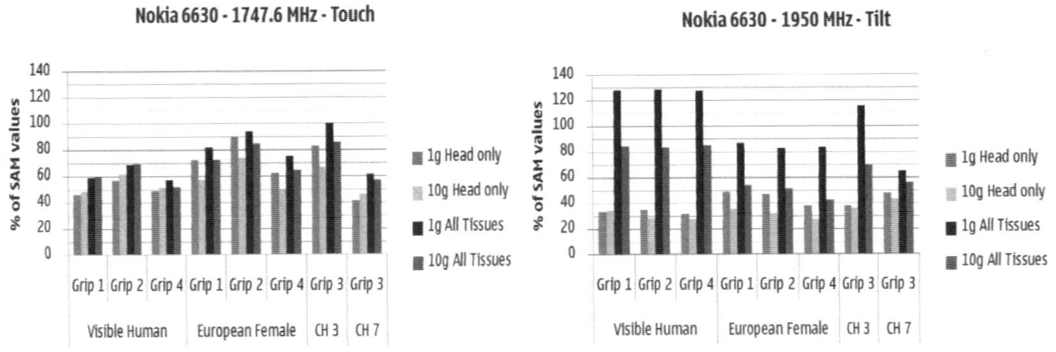

Figure 3. Relative SAR values for Nokia 6630 model in touch position for GSM upper band (left) and tilt position UMTS (right).

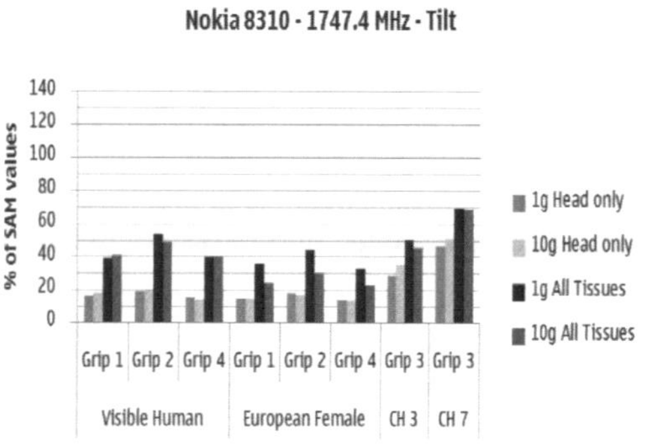

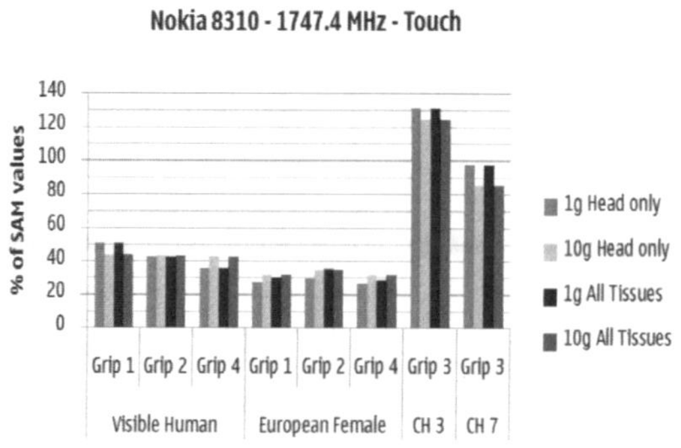

Figure 4. Relative SAR values for Nokia 6630 for tilt (left) and Nokia 8310 for touch (right) at 900 MHz

Analyzing the results following conclusions can be made:

1. In general the presence of the hand leads to significantly less conservative SAR values in the head for large majority of cases.
2. For lower band GSM frequencies the presence of the hand decreases the head SAR up to ~70%.
3. For the upper band GSM frequencies the presence of the hand decreases the head SAR up to ~55%.
4. The simulation results for Nokia 8310 for touch position at upper band GSM frequency for child head models which show significantly higher values are possibly due to the fact that in simulations the fingers intersect with the head tissues. This has to be further studied to identify the overlapping of fingers touching the face and its effect on head SAR values.
5. Based on the results of this study, the present SAR compliance protocol where hand phantom is not included leads to more conservative head SAR results compared to the cases where hand is included in the measurements.
6. The conclusions are based on this limited number of phone models, so more studies may be needed to make a firm conclusion.

5. References

[1] Li C H, Douglas M, Ofli E, Derat D, Gabriel S, Chavannes N, and Kuster N 2012 Influence of the hand on the specific absorption rate in the head IEEE Trans. Antennas Propagation vol 60 no 2 pp 1066–1074.

[2] Li C H, Ofli E, Chavannes N, and Kuster N 2009 Effects of hand phantom on mobile phone antenna performance IEEE Trans. Antennas Propagation vol 57 no 9 pp 2763–2770.

[3] *Human Exposure to Radio Frequency Fields from Handheld and Body- Mounted Wireless Communication Devices—Human Model Instrumentation and Procedures Part 1: Procedure to Determine the Specific Absorption Rate (SAR) for Hand-Held Devices Used in Close Proximity to the Ear (Frequency Range of 300 MHz to 3 GHz)* 2005 IEC Standard IEC62209-1.

[4] Pelosi M, Franek O, Knudsen M B, Pedersen G F and Andersen J B 2010 Antenna Proximity Effects for Talk and Data Modes in Mobile Phones IEEE Antennas and Propagation Magazine vol 52 no 3.

[5] *Protocol for the Computational Comparison of the SAM Phantom to Anatomically Correct Model of the Human Head at Extended Frequencies: 300 MHz – 5.8 GHz* 2008 IEEE SCC 34WG2.

[6] *IEEE recommended practice for determining the peak spatial-average specific absorption rate (SAR) in the human head from wireless communications devices: Measurement techniques IEEE Standard* 2003 IEEE SCC 34.

SAROTA: application of specific absorption rate (SAR) and over-the-air (OTA) data for the characterization of the real-life exposure due to mobile phones

Vikass Monebhurrun
Department of Electromagnetics, DRE-L2S, SUPELEC
3 rue Joliot-Curie 91192 Gif-sur-Yvette Cedex France

E-mail: vikass.monebhurrun@supelec.fr

Abstract. The RF exposure level of a mobile phone is quantified by the measurement of the specific absorption rate (SAR) under laboratory conditions. The SAR which is measured while the mobile phone is operated at maximum power level does not reflect the real-life exposure scenario since the mobile phone typically re-adjusts its power level and frequency depending on the quality of the communication link with the nearest base station. The choice of a low RF exposure device based on the comparison of the relative SAR values of mobile phones can be misleading. The real-life RF exposure also depends on the over-the-air (OTA) performance of the mobile phone. Taken independently, the two sets of data do not allow a straightforward comparison of the global RF performance amongst mobile phones. A unique and simple parameter denoted as the SAROTA index is proposed for the characterization of mobile phones with regard to both RF exposure and OTA performance. The SAROTA index provides the real-life exposure index of the mobile phone.

1. Introduction

The RF exposure level of a mobile phone is currently quantified by the measurement of the specific absorption rate (SAR) averaged over a given mass of tissue. As required by international standards [1], the SAR is measured in the laboratory using a standard dosimetric test facility and the specific anthropomorphic mannequin (SAM) while the mobile phone is operated at the maximum power level using a base station emulator. De facto, the SAR value does not reflect the real-life RF exposure of the mobile phone since in real situations it will typically re-adjust its output power level as well as the communication frequency depending on the radio link conditions with the nearest base station antenna. In Europe mobile phones are considered to be compliant with regard to RF exposure when the peak 10 g average SAR value does not exceed 2 W/kg. This limit is based on well-established thermal effects of RF exposure below which it is considered that the temperature rise in the body cannot be measured. Since the general public is increasingly concerned about RF exposure, it may be tempting to select a mobile phone based on the comparison of the relative SAR values or indices, e.g. select the mobile phone with the lowest SAR value after all other criteria such as aesthetics, design, etc. of the desired mobile phone are fulfilled. This mode of selection is misleading because the mobile phone with the lowest SAR value will not necessarily produce the lowest RF exposure in real-life situations. Indeed the over-the-air (OTA) performance of the mobile phone is also an important criterion. Clearly, a mobile phone with a badly designed antenna will typically provide a poor OTA performance –eventually compensated by RF emission at relatively higher power levels than may be required by a mobile phone with a better antenna design– as well as a low SAR value –that may be wrongly considered synonymous of a low RF exposure device. A test plan for the OTA characterization of mobile phones both with and without the SAM phantom has been already proposed by the Cellular Telecommunications and Internet Association (CTIA) [2]. However the exhaustive OTA measurement parameters –transmitter performance or total radiated power (TRP) and receiver performance or total isotropic sensitivity (TIS)– for each test configuration cannot obviously be provided to the untrained end-user. A unique and simpler parameter referred to as the SAROTA (SAR and OTA) index is proposed to characterize mobile phones with regard to both RF exposure and OTA performance [3]. The proposed index provides a better quantity for the comparison of the global RF performance amongst mobile phones and it additionally provides a real-life exposure index.

2. SAR compliance

The SAR measurement is typically performed inside a shielded anechoic chamber as shown in Figure 1. A six-axis robot is herein employed to scan for the electric field inside the SAM phantom filled with the appropriate tissue equivalent liquid. A base station emulator is employed to operate the mobile phone at the maximum power level. The SAR measurements are first performed for four intended use positions of the mobile phone against the SAM phantom –left/cheek, left/tilt, right/cheek and right/tilt– at the center frequency or traffic channel of the considered band, followed by two additional measurements carried out for the two extreme frequencies of the band for the worst-case phone/phantom position. This procedure is applied for all the operating frequency bands of the mobile phone and the SAR value ultimately provided to the end-user corresponds to the overall maximum value. Figures 2(a) and (b) show the results of the peak 10 g average SAR values of eleven dual-band mobile phones –numbered 1 to 11– obtained for the right/cheek position for the center frequencies of the Global System for Mobile (GSM) communication and Digital Communication System (DCS) frequency bands, respectively. Based on these SAR results, Phone No 7 would definitely be considered as a relatively higher RF exposure device, especially in the DCS frequency band. Based on SAR considerations, Phone No 10 would have been selected instead of Phone No 11 for both GSM and DCS frequency bands.

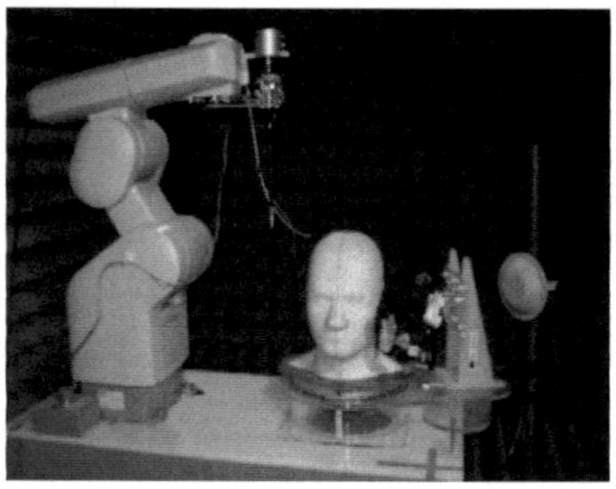

Figure 1. SAR measurement using a six-axis robot and the SAM phantom filled with tissue equivalent liquid. The SAR measurements are performed at the center frequency of a given band for four intended use positions of the mobile phone against the phantom. Two additional measurements are then performed at the two extreme frequencies of the band for the worst-case position.

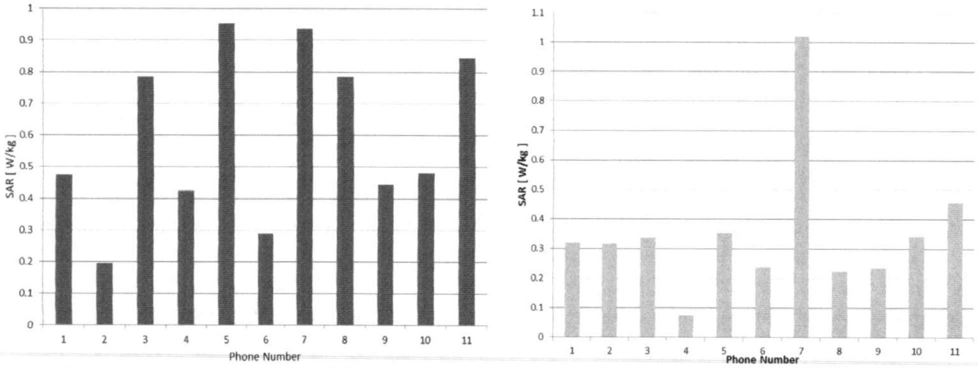

Figure 2. SAR values of eleven mobile phones for the right/cheek position: (a) GSM frequency band (left) and (b) DCS frequency band (right).

IOP Conf. Series: Materials Science and Engineering **44** (2013) 012018 doi:10.1088/1757-899X/44/1/012018

3. OTA performance

For the TRP measurements, the CTIA recommends the use of the traditional anechoic chamber antenna test facility. The mode-stirred reverberation chamber which is a good alternative for radiated power measurements of wireless communication devices is herein employed for the TRP measurements of the eleven mobile phones for the same configurations as those required for the SAR measurements i.e. same position and frequency (see Figure 3) [4]. Figures 4(a) and (b) show the results of the TRP values of the eleven mobile phones obtained for the right/cheek position for the center frequencies of the GSM and DCS frequency bands, respectively. Based on these TRP results, Phone No 2 would be considered as a relatively poor radiating device in the GSM frequency band. It is worth to note that this phone also showed a relatively low SAR value in the same frequency band. Based on TRP considerations, Phone No 11 would have been selected instead of Phone No 10 for both GSM and DCS frequency bands.

Figure 3. TRP measurement using a compact mode-stirred reverberation chamber. The TRP are measured for the same configurations (same frequency channel and use position) as those required for the SAR compliance of mobile phones.

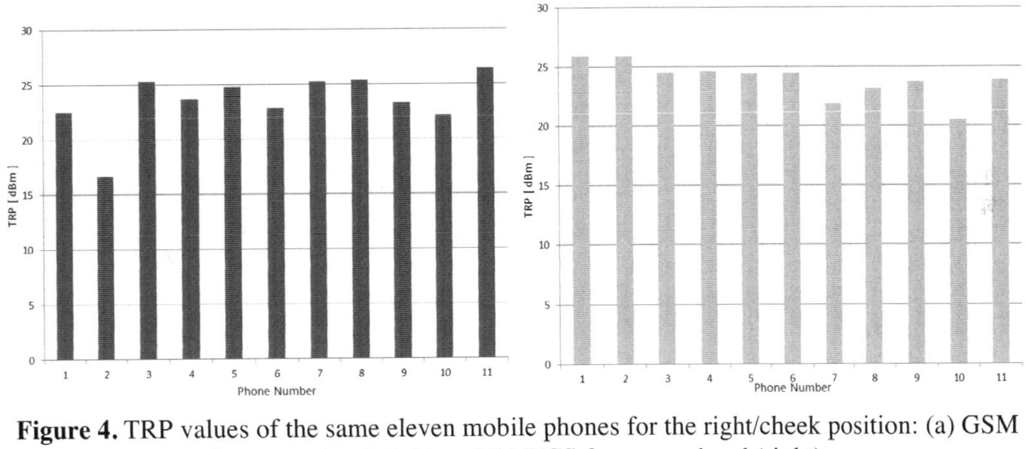

Figure 4. TRP values of the same eleven mobile phones for the right/cheek position: (a) GSM frequency band (left) and (b) DCS frequency band (right).

4. Real-life exposure: SAROTA index

Taken independently, the above sets of SAR and TRP data do not provide a straightforward indication about the mobile phone which offers the best compromise between RF exposure and OTA performance. The SAROTA index defined as the ratio between the SAR and the TRP values measured in the laboratory conditions for the same configuration, i.e. same maximum output power level, same position with respect to the head and same frequency channel, provides a simpler and easier quantification of both RF exposure and OTA performance. SAROTA is a measure of the power absorbed per kilogram of tissue per available power for communication, with units given in $W_{absorbed}/kg/W_{available}$:

$$SAROTA = \frac{SAR}{TRP} \tag{1}$$

Figures 5(a) and (b) show the SAROTA indices of the eleven mobile phones obtained for the right/cheek position for the center frequencies of the GSM and DCS frequency bands, respectively. Based on these indices, Phone No 4 provides relatively lower SAROTA values both for the GSM and DCS frequency bands. It is interesting to note that Phone No 2 provides the worst SAROTA index in the GSM frequency band. Furthermore, Phone No 7 provides relatively high SAR and low TRP values in the DCS frequency band which are reflected by a high SAROTA index. As expected, Phone No 11 offers the best SAROTA index as compared to Phone No 10, although the SAR index appeared to be favorable to Phone No 10. In real-life situation, if at a given location Phone No 10 has to emit at maximum power level in the GSM frequency band (i.e. about 23 dBm) to ensure a good communication link with the nearest base station, on average Phone No 11 will also emit at the same power level, i.e. the power will be lowered by about 3 dB with respect to the maximum that was measured in laboratory conditions (i.e. about 26 dBm). Consequently, the actual RF exposure or real-life SAR of Phone No 11 will be practically divided by two and found to be relatively lower than Phone No 10.

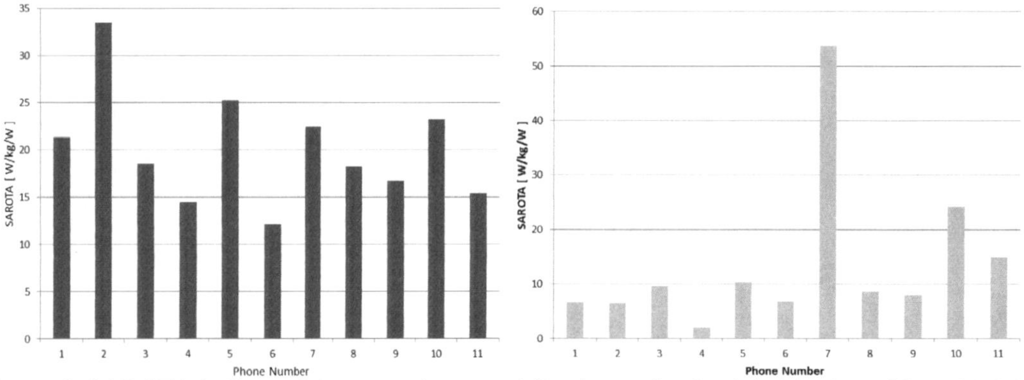

Figure 5. SAROTA indices of the same eleven mobile phones for the right/cheek position: (a) GSM frequency band (left) and (b) DCS frequency band (right).

5. Conclusion

Since SAR is measured under laboratory conditions while the mobile phone is operated at maximum power level, a comparison of the relative SAR values of mobile phones does not provide any indication about the real-life exposure. The latter depends on the OTA performance of the mobile. By taking into account both the SAR and OTA data measured under laboratory conditions, a real-life exposure index denoted as SAROTA is proposed. The SAROTA index is the ratio between the SAR and TRP values measured for the same configuration. It provides a simple quantity which aims to characterize a mobile phone with regard to both RF exposure and OTA performance.

6. References

[1] IEC Standard 62209-1 2005 *Human exposure to radio frequency fields from hand-held and body-mounted wireless communication devices – Human models, instrumentation and procedures – Part 1: Procedure to determine the specific absorption rate (SAR) for hand-held devices used in close proximity to the ear (frequency range of 300 MHz to 3 GHz)*

[2] CTIA Certification Test Plan for Mobile Station Over The Air Performance 2011 *Method of measurement for radiated RF power and receiver performance* Rev. 3.1

[3] Monebhurrun V 2010 A real-life exposure evaluation index for mobile phones based on SAR and TRP measurements data: the SAROTA index *Proc. 32nd Annual Meeting of the Bioelectromagnetics Society (Seoul)* p 1-2

[4] Monebhurrun V 2007 Fast global characterization of 2G and 3G mobile phones using a reverberation chamber *Proc. 2nd European Conference on Antennas and Propagation (Edinburgh)* p 1-7

Radio and Antenna Days of the Indian Ocean (RADIO 2012) IOP Publishing
IOP Conf. Series: Materials Science and Engineering **44** (2013) 012019 doi:10.1088/1757-899X/44/1/012019

GALFACTS: The Galactic ALFA Continuum Transit Survey

A.R. Taylor

Department of Physics and Astronomy, University of Calgary, 2500 University Dr. N.W., Calgary, Alberta, Canada, T2N 1N4

E-mail: russ@ras.ucalgary.ca

Abstract. The Arecibo Radio Telescope has been equipped with a seven feed horn array operating at 1400 MHz - the Arecibo L-band Feed Array (ALFA). This new feed system, combined with a 300 MHz digital spectrometer back end, enables the large collecting area of the Arecibo reflector to be utilized for sensitive, wide-area imaging. A consortium of over 40 researchers from around the world is working together to use this system to carry out a spectro-polarimetric imaging survey of the entire sky visible to the Arecibo Telescope; the Galactic ALFA Continuum Transit Survey (GALFACTS). GALFACTS covers 13,000 square degrees of sky, creating over this area full-Stokes image cubes at angular resolution of $3.5'$ with several thousand spectral channels covering 1225 to 1525 MHz, and band-averaged sensitivity of 90 microJy. The complete survey will require about 2000 hours of observing time. Observations began in 2009 and are now 80% complete. The data will provide a rich new database for exploration of the magnetic field of the Galaxy, the properties of the magneto-ionic medium and the polarization properties of over 10^5 extragalactic radio sources.

1. Introduction

The Arecibo 305-m antenna is the world's largest single-dish telescope, achieving continuous spatial frequency coverage, high brightness sensitivity, and arcminute-scale angular resolution at decimetre wavelengths. The new ALFA multi-beam receiver system allows these properties to be used to efficiently image large areas of the sky at $\lambda 21$ cm. The GALFACTS Consortium is using the Arecibo telescope and ALFA to carry out a sensitive, high resolution, spectro-polarimetric survey of the region of the sky visible with the Arecibo telescope – the GALFA Continuum Transit Survey, GALFACTS.

A key observational objective of the GALFACTS is to image the polarized emission from both discrete objects and the diffuse interstellar medium of our Galaxy and to derive polarization properties, including Faraday Rotation Measures for a vast population of extragalactic sources. Wide-field polarimetric imaging of Galactic emission has enjoyed a surge of activity over the past decade with the detection of highly-structured, arcminute-scale polarized radiation in interferometric images, both at high Galactic latitude, e.g. in WSRT 349-MHz images (Haverkorn 2003), and in the plane of the Galaxy with the Canadian Galactic Plane Survey (Taylor et al. 2003, Landecker et al. 2010) and Southern Galactic Plane Survey (Haverkorn et al. 2006) at 1.4 GHz. Similar structures are seen at 5 GHz by (Sun et al. 2007). These structures are superposed on the polarized emission from the diffuse Galactic synchrotron emission, but themselves have no Stokes-I counterpart. The accepted interpretation is that the distributed polarized emission arises from the intrinsically-smooth Galactic synchrotron emission, but differential Faraday rotation in the intervening magneto-ionic medium (the Faraday Screen) imposes fine structure due to the resulting spatial variation in polarization position angle. Propagation effects in the interstellar medium dominate over intrinsic polarized emission structure. Observations of diffuse polarized emission and of the propagation effects of the ISM on the Rotation Measures of background sources are thus powerful probes of both the relativistic and thermal plasmas of the interstellar medium.

GALFACTS is a major observational advance by virtue of its high brightness sensitivity compared to interferometric surveys, the high angular resolution compared to other single-dish telescopes and the large spectral bandwidth, which combined with high spectral resolution will allow accurate measurement of the frequency dependence of the polarization state of the emission at 1.4 GHz. Presently, the highest resolution, single-dish, L-band continuum surveys are those of the northern sky by the Effelsberg telescope (Reich et al. 1997) and in polarization by (Uyaniker et al. 1999) of a few selected regions at intermediate latitudes, both with resolution of $9.4'$ and brightness sensitivities of 10's of mK. For surveys covering a significant fraction of the sky in polarization, only significantly poorer resolution data are available; $\theta_{\rm HPBW} = 35'$, (Wolleben et al. 2006) at similar sensitivity (12 mk). In comparison, GALFACTS will provide full-Stokes image cubes at resolution of $3.5'$, at a brightness sensitivity an order of magnitude deeper than previous work, and covering a frequency band from $1225 - 1525$ MHz in spectral-line mode, allowing imaging of Faraday Rotation Measure.

2. Observations

The observations are carried out with the Arecibo L-band feed array system, ALFA. The seven feed horns of the ALFA provide seven beams on the sky with six beams arranged in a hexagonal pattern around a central on-axis beam. Each feed horn detects two nominally orthogonal states of linear polarization. Data is acquired simultaneously in each polarization channel while the feed system is alternately scanned north and south along the meridian at a rate of 1.53 degrees per sidereal minute. Combined with the Earth's rotation, this tracking speed creates a zig-zag track pattern on the sky with the tracks from individual beams separated by $1.83'$, close to Nyquist sampling for the $3.5'$ FWHM beams. The tracking geometry is illustrated in Figure 1. On consecutive days the track of the central beam is shifted by 51 seconds of Right Ascension. For 18 degree long Declination scan, 28 separate days/tracks provides complete coverage.

3. Data Acquisition and Processing

The FPGA spectrometer backend (the Mock spectrometers) produces data streams for each of four polarization states and 4096 spectral channels for all seven beams in two adjacent 172-MHz bands, which together cover the 300 MHz total band from 1225-1525 MHz. During the observations a continuous square-wave noise diode signal is injected into the two polarization channels at the feed horns at 25 Hz with 50% duty cycle to calibrate the time dependence of the complex gains of the receiver system. The noise diode is injected with equal strengths into the X and Y polarization channels. Each of the fifty-six (seven beams, four polarizations, two bands) 4096-channel data streams are sampled at 1 millisecond per sample to allow the on and off states of the noise diode to be detected in post processing. The aggregate data rate is 460 MB/s.

Data are recorded directly onto disk drives at Arecibo Observatory. After each observing day the data are run through first-stage processing which separates the on and off states of the noise diode signal and time averages the data to 0.2 s sample time diode-on and diode-off data streams. These time-averaged data are shipped to the University of Calgary at the end of each observing run for further processing.

IOP Conf. Series: Materials Science and Engineering **44** (2013) 012019 doi:10.1088/1757-899X/44/1/012019

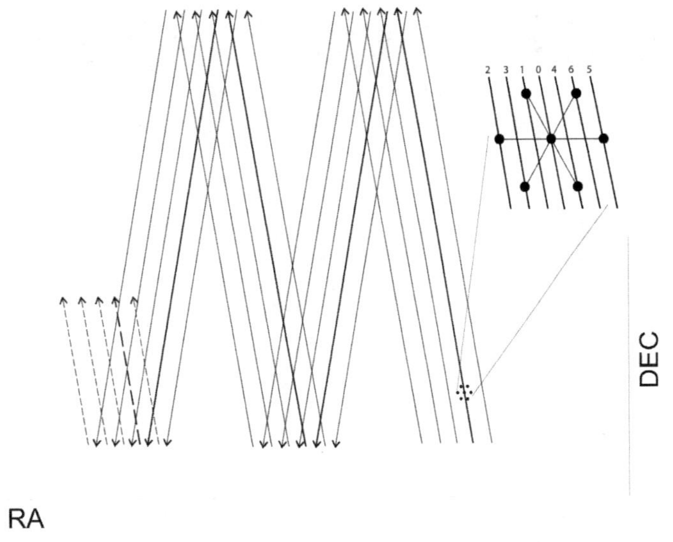

Figure 1. GALFACTS observations are made by scanning the seven-feed ALFA receiver up and down the meridian at a rate of 1.53 degrees per sidereal minute. Each day of observations creates a zig-zag patterned set of up and down "scans", each consisting of seven sub-tracks (see inset at upper right), one sub-track from each of the ALFA beams. Main tracks for successive days of observations are separate by 51 seconds of Right Ascension. Over 28 days a fully-sampled strip of sky is covered that is 18 degrees wide in Declination over 6 hours of Right Ascension.

Calibration of the data is simplified by the fact that we observe only along the meridian at a fixed feed rotation. There are thus only three independent variables for instrumental effects, frequency, time and elevation angle. The data from each day and beam are corrected for time variations of the receiver channel gains using the noise diode signals. This process converts the raw data into Stokes parameters in units of brightness temperature. Frequency-dependent baselines are remove by fitting high order polynomials to the elevation dependence of the signal amplitude in regions removed from strong sources for each day, beam and frequency channel. The Stokes I bandpass shape and the polarization leakage corrections derived from calibration observations are then applied and the corrected data are "basketweaved" to align total intensity offsets between days and gridded into images. The frequency-dependent baseline subtraction removes the emission on spatial scales of order 10 degrees and longer. These missing spatial frequencies will be recovered by integrating data from the Global Magneto-ion Medium Survey, GMIMS (Wolleben et al. 2009). Further processing, including cleaning the dirty images to remove sidelobes from the multiple beams (Guram 2008) is carried out on the gridded image data.

The final data products from GALFACTS are spectro-polarimetric image cubes, (intensity as a function of α, δ, ν), in each Stokes parameter I, Q, U and V, with $\sim$8000 spectral channels over a 300 MHz bandwidth. For a 3.5$'$ FWHM beamsize the effective integration time per scan for a given direction on the sky is 2.4s, or a total integration time per beam area of 4.8s. For an average SEFD over the seven beams of 3.4 Jy, the theoretical noise level per 42 kHz channel is 7.5 mJy per polarization. Integrated over the full 300-MHz band the theoretical 1σ point-source sensitivity level is 90 μJy per polarization, or a surface brightness sensitivity of 1 mK.

A preliminary set of GALFACTS images is shown in Figure 2. These images show total intensity as well as polarized emission in Stokes parameters Q an U over a 170 MHz bandwidth

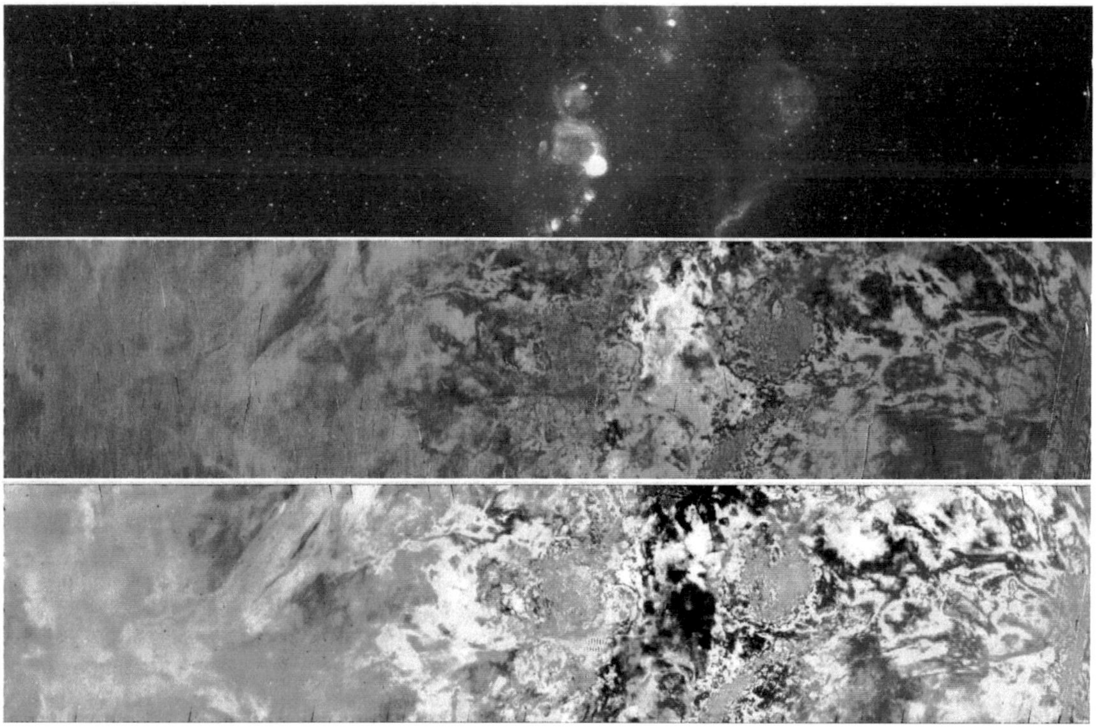

Figure 2. Sample GALFACTS images of one 1700 square degree sub-region covering 18 degrees of Declination an 6 hours of Right Ascension. The top panel shows the total intensity and the middle and bottom panels shows polarized emission in Stokes Q and U respectively.

at 1440 MHz. The band of highly-structured emission running vertically through the total intensity image is the Galactic plane, containing the Rosette Nebula and λ Orionis. The polarized emission is structured throughout, primarily due to spatial variations in polarization position angle from propagation of Galactic synchrotron radiation through the Faraday screen of the Galactic magneto-ionic medium. As one moves away from the plane the polarization structures persist but become smoother and larger scale. The magneto-ionic effect of the λ Orionis bubble is clearly seen in polarized intensity, defined by a smooth circular region internal to the bubble, surrounded by a ridge of polarized emission.

References

[1] Haverkorn M, Katgert P and de Bruyn A B G 2003 *A & A Supp.* **126** 413
[2] Taylor, A R, et al. 2003 *Astronomical Journal* **125** 3145
[3] Landecker, T L, et al. 2010 *A & A* **1004** 2536
[4] Haverkorn M, Gaensler B M, McClure-Griffiths N M, Dickey, J M and Green A J 2006 *Astrophysical Journal* **167** 230
[5] Sun X H, Han J L, Reich W, Reich P, Shi W B, Wielebinski R and Fürst E 2007 *A & A* **463** 993
[6] Reich P, Reich W and Fürst E 1997 *A & A Supp.* **126** 413
[7] Uyaniker B, Fürst E, Reich W, Reich P and Weilebinski R 1999 *A & A Supp.* **138** 31
[8] Wolleben M, Landecker T L, Reich W and Wielebinski R 2006 *A & A* **448** 411
[9] Wolleben M, et al. 2009 in *Cosmic Magnetic Fields: From Planet, to Stars and Galaxies*, ed K G Strassmeier, A G Kosovichev and J E Beckman (UK: Cambridge University Press) **259** of IAU Symp 89
[10] Guram, S 2008, M.Sc. Thesis, University of Calgary

IOP Conf. Series: Materials Science and Engineering **44** (2013) 012020 doi:10.1088/1757-899X/44/1/012020

Expanding radio astronomy in Africa

M J Gaylard

HartRAO, P. O. Box 443, Krugersdorp 1740, South Africa

E-mail: mike@hartrao.ac.za

Abstract.
The Square Kilometre Array (SKA) Organisation announced in May 2012 that its members had agreed on a dual site solution for the SKA [1]. South Africa's bid for hosting the SKA has caused a ramp up of radio astronomy in Africa. To develop technology towards the SKA, the South African SKA Project (SKA SA) built a protoype radio telescope in 2007, followed in 2010 the seven antenna Karoo Array Telescope (KAT-7). Next is the 64 antenna MeerKAT, which will merge into SKA Phase 1 in Africa.

As SKA Phase 2 is intended to add a high resolution capability with baselines out to 3000 km, the SKA SA brought in partner countries in Africa to host outstations. South Africa has been working with the partners to build capacity to operate the SKA and to benefit from it. The SA Department of Science and Technology (DST) developed a proposal to establish radio telescopes in the partner countries to provide hands-on learning and a capability for Very Long Baseline Interferometry (VLBI) research. Redundant 30 m class satellite antennas are being incorporated in this project.

1. The SKA in Africa

Radio astronomy in Africa is set for radical change with the announcement on 25 May 2012 of dual siting for the SKA in Southern Africa and Australia. The majority of SKA Phase 1 dishes will be built in a ~100 km radius around the core site near Carnarvon in the arid Karoo region of South Africa, and will incorporate MeerKAT, comprising 64 13.5-m dishes. The balance of the dishes will be built in Australia to work with the ASKAP array of 36 12-m antennas. All the dishes and mid-frequency aperture arrays for Phase 2 will then be built in Africa. The Phase II dishes will add to the core, and also provide outstations on baselines to 3000 km. In Phase 1 the dishes will operate from 0.45 to 3 GHz and in Phase 2 this will be extended to 10 GHz.

2. SKA precursors in Africa - XDM, KAT-7 and MeerKAT

In addition to bidding for siting the SKA, SKA SA also worked on developing technologies for the project, including reflector antennas to be suitable for mass production at low cost. The first prototype antenna was a 15-m, prime focus, symmetric paraboloid called the "eXperimental Development Model" (XDM). This was constructed at the existing Hartebeesthoek Radio Astronomy Observatory (HartRAO), west of Johannesburg, in 2007. An accurate mould of the dish shape was assembled on site. A releasant layer was sprayed on it, followed by a flame-sprayed aluminium layer. Foam blocks wrapped in fibreglass were placed on top and resin sucked in to create a one-piece dish. The steel backing frame was bonded to this. Two receivers for 1.4 - 1.7 GHz were fitted. After completion of testing HartRAO re-equipped the telescope with a dual-band cryogenic receiver for 2.3 and 8.4 GHz, intended primarily for geodetic VLBI.

This was followed by the Karoo Array Telescope KAT-7 array of seven 12-m antennas, again of prime-focus symmetric paraboloid design. They are equipped with single 1.4 - 1.8-GHz cryogenic receivers with Stirling coolers taking them to liquid nitrogen temperatures (77 K). These were built using an improved version of the process employed for XDM, and were constructed on the plain near Carnarvon where the SKA will be built. The maximum baseline is 185 m. Completed in 2010, the KAT team has since been commissioning an expanding range of observing modes. Some science research has been carried out during this phase.

The next iteration in development is MeerKAT (more KAT), which is designed to provide a competitive science instrument. It will comprise 64 offset Gregorian antennas of 13.5 m effective diameter. Three receivers will cover 0.58 - 1.015 GHz, 0.9 - 1.76 GHz and 8 - 14 GHz. Array baselines will be out to 10 - 20 km. The winning bid for the antennas provided conventional construction of aluminium panels on steel structure. Large scale survey projects have already been accepted for MeerKAT, but it is also to be integrated into SKA Phase 1.

3. HartRAO and long baseline astronomy

The 26-m antenna at the Hartebeesthoek Radio Astronomy Observatory (HartRAO) was built as a NASA spacecraft tracking antenna in 1961 and it was operated for NASA until 1974. Since then it has operated as a South African National Facility for radio astronomy. The antenna has been subject to many upgrades and is currently equipped with seven receivers. Its primary science function today is for VLBI, but it also used for research operating as a single dish. For astronomical VLBI it operates with the European VLBI Network (EVN) and Australia Telescope Long Baseline Array (AT-LBA), and for astrometric and geodetic VLBI it operates with the International VLBI Service for Geodesy and Astrometry (IVS). Its value lies in providing long baselines to other continents, but for astronomical imaging the drawback is that there are no other VLBI-capable telescopes in Africa or adjacent islands to fill the long gaps to Europe and Australia. This means images cannot be unambiguously created from the data.

4. Radio astronomy development in partner countries

South Africa's DST and SKA SA identified the need to provide training for the African partner countries to maximise the benefit of the SKA for Africa. The training needs are both scientific, to be part of the exploitation of SKA's research capabilities, and technical, to be part of the operation of the SKA outstations. DST developed a proposal for relatively modest telescopes to be established in partner countries, which would also have a VLBI capability.

5. Satellite Earth stations and fibre-optic communication

Communication via geostationary satellites was introduced to carry voice, data and TV signals, using the 5.9 - 6.4 GHz band for uplink and 3.7 - 4.2 GHz band for downlink. About 30 Satellite Earth Stations built in Africa from 1970 to 1985 were equipped with Intelsat Standard A antennas of 30 - 32-m diameter. New wideband fibre optic links have made the large antennas redundant. The DST proposal has incorporated the conversion of redundant large antennas to radio telescopes. Their location in partner countries is shown on the SKA SA developed layout for SKA outstations shown in Fig. 1. The feasibility of the proposed conversions is indicated by the success of previous conversions, at Ceduna in Australia and Yamaguchi and Ibaraki in Japan. New Zealand, Peru, Ireland and England are also carrying out conversions. The 34-m NASA DSS-28 beam waveguide antenna in California has also been converted for radio astronomy [2].

6. Astronomical applications for large antennas in Africa

For astronomical imaging with existing VLBI networks it would be very helpful to have antennas filling the gap between SA and Europe and SA and Australia. To create an independent VLBI

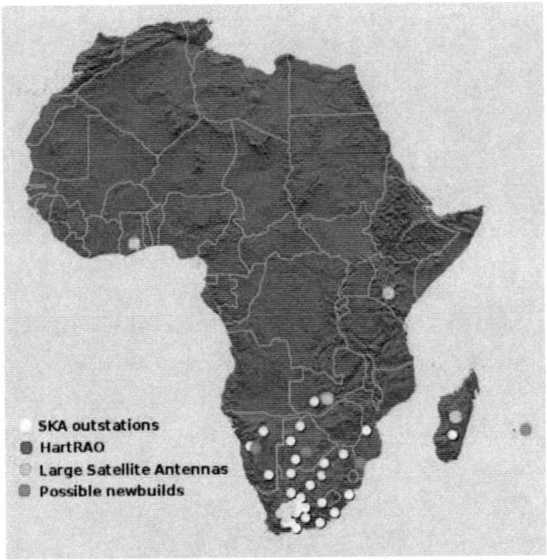

Figure 1. Approximate locations of outstations proposed by SKA SA, existing large antennas in partner countries, and possible new build antennas.

network able to produce images, a minimum of four antennas are needed to provide both phase and amplitude closure.

Development of the science case is guided by usage of existing VLBI networks and by the science case developed for the SKA [3], [4]. VLBI is used to observe radio-bright sources of small angular size. Large objects physically meet this requirement if they are distant. Active galactic nuclei and radio-bright supernovae in external galaxies are examples of objects whose evolution can be studied with VLBI. Masers in star-forming regions in the Milky Way are examples of nearby bright, compact sources. Methanol masers at 6.668 GHz and 12.178 GHz are of particular interest currently. Measurement of their annual parallaxes by repeated VLBI observations enables their distances to be determined, and thus also the locations of the spiral arms in the Milky Way. The same technique can be used to measure pulsar distances. Transient sources and gamma-ray bursters are both potential targets.

Antennas equipped with 2.3+8.4 GHz receivers could participate in geodetic and astrometric VLBI with the current generation of radio telescopes carrying out this research. This technique provides very precise absolute location for the telescope, to which co-located relative position measuring systems such as GPS and satellite orbit measuring systems can be tied.

VLBI is not expected to run continuously on African telescopes. The time available for single-dish astronomy would be valuable for student training purposes and for selected research projects. Radiometry, spectroscopy and pulsar timing capabilities would be established.

7. Towards an African VLBI Network - conversion of existing antennas

The Ghanaian government has taken over the redundant 32 m antenna at the Kutunse Satellite Station outside Accra from Vodafone. SKA SA and HartRAO have built up an engineering team to assist the Ghana Space Science and Technology Centre (GSSTC) with its conversion. This antenna is well placed to fill the gap between Europe and Africa for VLBI with the EVN. The aim is to establish an initial operational capability in the 5 GHz VLBI band. Discussions are also underway with Kenya and Zambia regarding their large antennas.

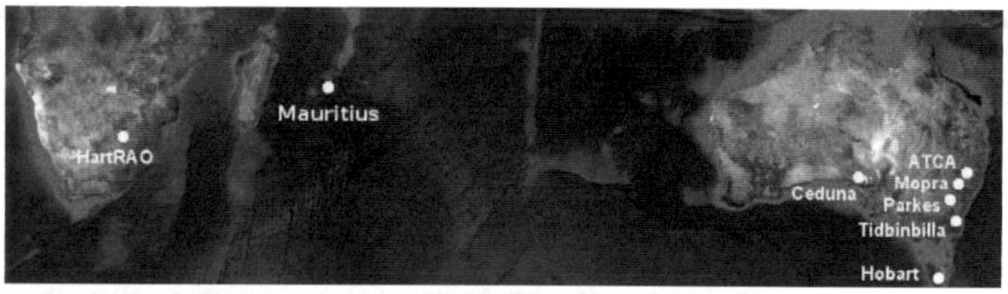

Figure 2. Location of Mauritius between African and Australian radio telescopes.

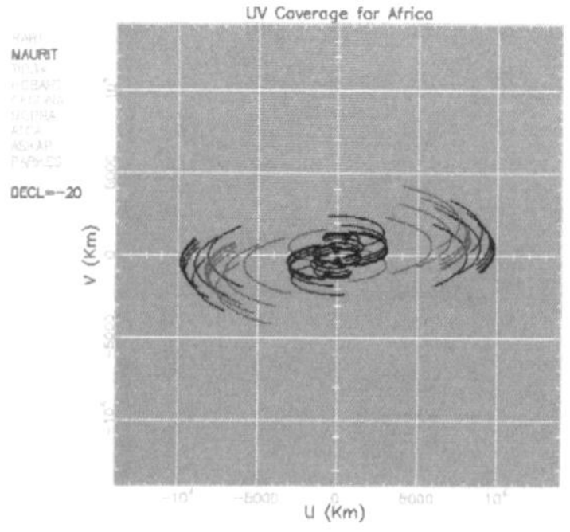

Figure 3. UV coverage for a southern source observed with the AT-LBA and HartRAO. The red tracks show the improved coverage provided by adding a radio telescope on Mauritius.

8. Towards an African VLBI Network - potential new telescopes

Four partner countries do not have large antennas, so the feasibility of establishing new radio telescopes for training and VLBI is being investigated. Mauritius has a history of radio astronomy using the 151 MHz Mauritius Radio Telescope (MRT), and so it has taken the lead in joining this study. The design and siting of potential new telescopes is driven by the science case, the need for compatibility with existing networks and converted antennas, the provision of services, minimisation of radio frequency interference, and of course, costs. A two day workshop was held on Mauritius on 21 and 22 September 2012 to introduce Mauritian role players to the subject and establish work groups. Mauritius is well situated to fill the gap between Africa and Australia, as shown in Fig. 2. The UV plot for a typical southern source is shown in Fig. 3.

9. Summary

The coming of the SKA is greatly expanding radio astronomy in Africa through multiple precursor projects. The potential for converting obsolete large antennas for radio astronomy has been recognised World-wide, and this is an opportunity for Africa too. The case for potential new-build telescopes is being developed with the participation of Mauritian radio astronomers.

References

[1] *SKA Members Statement on the General Meeting of 25 May 2012* http://http://www.skatelescope.org/news-/ska-members-statement-general-meeting-25-2012/
[2] Imbriale A A, Weinreb S, Jones G, Mani H and Akgiray A 2011 *IEEE Trans. Ant. Propagat.* **59** 1954
[3] Godfrey L, Bignall H and Tingay S 2011 *Memo 135 Very High Angular Resolution Science with the SKA*, http://www.skatelescope.org/publications
[4] Godfrey L E H, Bignall H, Tingay S, Harvey-Smith L, Kramer M, Burke-Spolaor S, Miller-Jones J C A, Johnston-Hollitt M, Ekers R and Gulyaev S 2012 *Publ. Astron. Soc. Australia* **29** 42

IOP Conf. Series: Materials Science and Engineering **44** (2013) 012021 doi:10.1088/1757-899X/44/1/012021

Engineering Processes for the African VLBI Network

Venkatasubramani L Thondikulam[1], Anita Loots[1] and Michael Gaylard[2]

[1] SKA-SA, The Park, Park Road, Pinelands 7405, South Africa
[2]HartRAO, PO Box 443, Krugersdorp 1740, South Africa

E-mail: venkat@ska.ac.za, anita@ska.ac.za, mike@hartrao.ac.za

Abstract. The African VLBI Network (AVN) is an initiative by the SKA-SA and HartRAO, business units of the National Research Foundation (NRF), Department of Science and Technology (DST), South Africa. The aim is to fill the existing gap of Very Long Baseline Interferometry (VLBI)-capable radio telescopes in the African continent by a combination of new build as well as conversion of large redundant telecommunication antennas through an Inter-Governmental collaborative programme in Science and Technology. The issue of human capital development in the Continent in the techniques of radio astronomy engineering and science is a strong force to drive the project and is expected to contribute significantly to the success of Square Kilometer Array (SKA) in the Continent.

1. The Motivation for AVN

The AVN is strongly motivated from scientific [1,2] and political perspectives.

The 26m radio telescope, HartRAO, at Hartebeesthoek in South Africa is the only VLBI-capable instrument currently operational in the entire African continent. The absence of a similar telescope between Africa and Europe as well as Africa and Australia for astronomical, astrometric and geodetic VLBI is a strongly felt scientific need for the past many decades. The AVN will be filling this need and will also act as a convenient platform for co-locating geodetic instrumentation and other scientific facilities in the same infrastructure.

With a major share of SKA having come to the African Continent, there is a strong political motivation to facilitate activities in the Science and Technology of Radio Astronomy in the SKA Partner Countries, well ahead of the SKA roll-out programme. AVN is seen as an immediate vehicle for this goal and act as a pathfinder towards SKA. It is managed by the Department of Science and Technology of South Africa, with active participation from partner countries, to address the issues of human capital development in the continent with a specific focus on specialized skills transfer to operate, maintain and sustain a radio astronomy facility.

SKA-SA and HartRAO have been mandated by DST, South Africa, to carry out the above task with a goal of VLBI Demonstration by 2017.

2. The Solution for AVN

The SKA in the African continent will encompass radio astronomy facilities in eight partner countries, namely Botswana, Ghana, Kenya, Madagascar, Mauritius, Mozambique, Namibia and Zambia with the core located in South Africa.

Some of these partner countries have large 30m diameter class redundant telecommunication antennas which could be converted as radio telescopes of the AVN at a moderate investment. Work along these lines has been initiated in Ghana and will be started shortly in Kenya. Other partner countries will need a new-build radio telescope with a performance acceptable by the International VLBI Community to be included as a part of the existing European VLBI Network (EVN) and other consortia. A nodal processing facility will be located in South Africa when about four of the VLBI stations are successfully operational and thus result in the African VLBI Network.

With the goal being clearly set as above, issues of political, scientific, technical and management nature are currently being addressed as under:

- Is there a strong inter-governmental foundation and support available for the effort to succeed?
- What is the most efficient way to do what we want to achieve?
- What are the initial tests and studies required?
- Who is responsible for what?
- How to make the Human Capital Development efforts successful?
- What resources are required (budget, manpower) and where do we get them?

3. The Groundwork towards a Technical Solution

The Implementation Flow Diagram for AVN under Conversion process is given in figure 1. It visualizes a Memorandum of Agreement with an Operating Institution with a clear top-down mandate as an important pre-requisite.

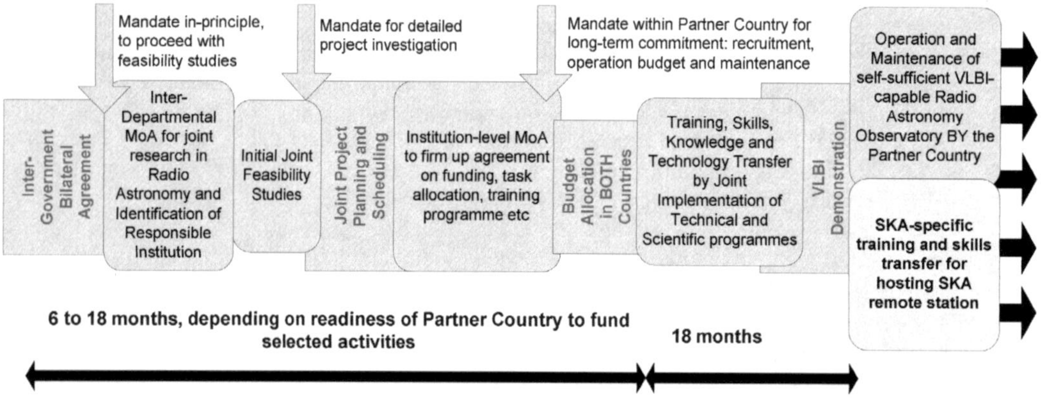

Figure 1 Implementation Flow Diagram for AVN - Conversion Process.

A similar model applies for new-build stations to produce a firm foundation for the effort. In this instance, the initial technical feasibility studies in Figure 1 will have to be aimed towards developing a decision matrix for optimal location of the station considering all aspects like Radio Frequency Interference (RFI) potential, approach roads, power, water, geo-technical and topographical studies, optical fibre connectivity, environmental factors (wind, humidity, temperature) and others.

A criterion to define the new-build VLBI station at such a location chosen as above in the form of initial frequency of operation and a figure of merit for sensitivity as used by VLBI community has evolved. *Considering the requirements of astronomical and geodetic VLBI as well as non-VLBI programmes of pulsar timing and spectroscopy, the frequency bands to support have been chosen as (a) 1.4-1.73 GHz; (b) 2.2-2.4 GHz; (c) 4.7-5.2 GHz; (d) 6.6-6.8 GHz and (e) 8.2-9.0 GHz.*

Table 1 gives the figure for System Equivalent Flux Density (SEFD), defined as the number of Janskys (10^{-26} W m^{-2} Hz^{-1}) of source flux density that are required to double the system noise temperature, for some of the important operating VLBI stations around the World. *From this table the target SEFD for the AVN at 5 GHz has been set at better than the median value of 220 Jy.*

This figure would translate into a decision matrix for the receiver system with diameter of each antenna, an array of antennas from 1 element to a reasonable number and choice of the receiver with a cryogenic or non-cryogenic front-end as parameters.

The decision matrix for site and system selection as described above will have to be optimized for "Expectation – Cost – Performance" trade-offs before embarking on realization.

IOP Conf. Series: Materials Science and Engineering 44 (2013) 012021 doi:10.1088/1757-899X/44/1/012021

Table 1 SEFD of some of the operating VLBI Stations.

Telescope	EVN Code	Diameter (m)	SEFD (1.6 GHz)	SEFD (5GHz)
EVN				
Jodrell Bank	JB-1	76	65	80
Jodrell Bank	JB-2	25	320	320
Cambridge	Cm	32	212	136
Westerbork (14)	Wb	25	330	760
Effelsberg	Eb	100	19	20
Medicina	Mc	32	700	170
Noto	Nt	32	784	260
Onsala	On-85	25	320	600
Shanghai	Sh	25	670	720
Urumqi	Ur	25	270	200
Torun	Tr	32	300	220
Yebes	Ts	40		160
Arecibo	Ar	305	3	5
Hartebeesthoek	Hh	26	430	730
Svetloe	Sv	32	360	250
Zelenchukskaya	Zc	32	300	400
Badary	Bd	32	330	200
MEDIAN SEFD (Jy)		**32**	**320**	**220**
VLBA (10)		25	303	210
AT-LBA				
ATCA (6)		22	282	420
Hobart		26	420	640
Ceduna		30		450

4. The Engineering Process towards Realisation

The System Engineering [3] approach, which has been successfully demonstrated for the SKA precursors in South Africa, will be adapted and adopted for AVN. This adaptation needs to take into account the way the Partner Country team works and the practicalities with respect to collaboration with South Africa, without jeopardising the integrity of the process.

The approach needs a process of Validation and Verification (V&V) at all levels of development and implementation as depicted in the V-Diagram in figure 2. The Requirements, as specified by the User (AVN URS) for New-Builds and Conversion approaches, generates a System Description Document (AVN SDD) as an output to a system design activity. It describes a high level solution to meet the requirements and enables the required functionality associated with each requirement to be defined. The System Engineer uses the AVN SDD as an input to produce a Station System Requirement Document (Station SRD) and an associated Station System Description Document (Station SDD), with traceability to the AVN URS. The SDDs will be used to derive all aspects of budget estimates, time schedules, resources etc in a System Engineering Management Plan (SEMP). There could be a few iterations as depicted, to result in a mature Station SDD which is acceptable to all stake-holders as a good “requirement – cost – time – performance” optimised solution.

The Station SDD will become the input to a team of Design Engineers to derive the next level of sub-system SRDs and associated SDDs. The typical sub-systems for a radio telescope will have Signal Chain (SC), Software (SW), Control and Monitor (CAM), Structural-Mechanical (SM) and Other (Infrastructure, Operations, maintenance, logistics etc) elements. Each of the sub-systems to form the telescope could have interfaces between any or all of the others which needs to be adequately and unambiguously captured in the Interface Control Document (ICD). Each of the sub-system SDDs will need to be addressed by the Development Engineer at Equipment level (ERD and EDD) as well as component level (CRD) to result in a low-level realisation of the building blocks of the telescope.

Associated with each SRD is a Qualification Test which is performed at the same level as where the SRD resides, e.g. a derived requirement appearing in the SRD for the azimuth angle decoding subsystem will have an associated SDD describing the solution and a Qualification Test Procedure (QTP) to verify that the SRD requirement has been met. (Right hand V in figure 2). This test will typically be done on a jig of the subsystem and performance verified before the higher level integration (installed onto telescope) can begin. Thus the documentation at various levels as the process percolated down from the User to Component will be used for verification and validation as the telescope takes shape.

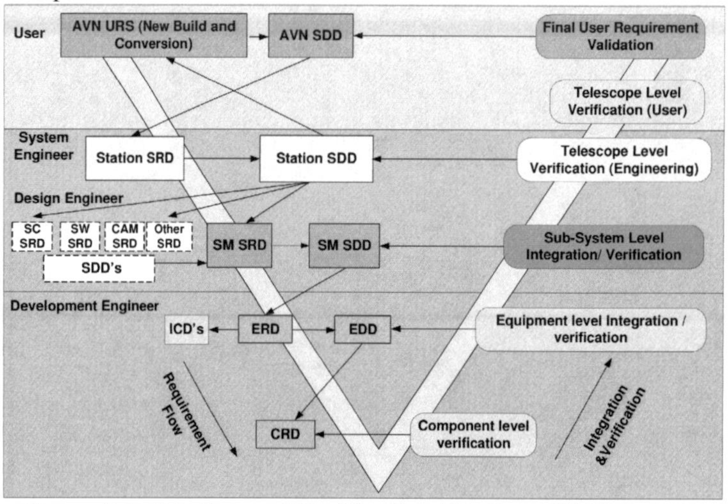

Figure 2 Validation and Verification Diagram (V-diagram).

The system engineering process as described above has been widely used in a rigorous manner in the Industry and is expected to contribute considerably to the success of AVN as it rolls out.

5. Human Capital Development Aspects in AVN

Human Capital Development with specialized skills transfer is a vital element for the success of AVN. To assist in this, a training programme in South Africa is being developed for core technical and scientific members from each of the partner countries. The aim of the initial training to a focused and dedicated technical team from each of the partner country will be for operation and maintenance of the telescope system in that country. The program is expected to enlarge with appropriate skill development in various disciplines as the work moves towards SKA.

6. Summary

SKA is acting as a motivating driver in facilitating setting up experimental facilities for radio astronomy research in Africa. The African VLBI Network, when realized, will cater to a long awaited scientific need in the Continent and will serve as a pathfinder towards SKA

References

[1] Gaylard M J, Bietenholz M F, Combrinck L, Booth R S, Buchner S J, Fanaroff B L, MacLeod G C, Nicolson G D, Quick J F H, Stronkhorst P and Venkatasubramani T L 2011 *An African VLBI Network of radio telescopes* Proccedings of SAIP2011, the 56th Annual Conference of the South African Institute of Physics, edited by I. Basson and A.E. Botha (University of South Africa, Pretoria, 2011), pp 473-478, ISBN: 978-1-86888-688-3. Available online at http:\\www.saip.org.za

[2] Gaylard M J 2012 *Expanding radio astronomy in Africa* Proceedings of the RADIO 2012 Conference, Mauritius

[3] INCOSE 2011 *"System Engineering Handbook – A Guide for System Life Cycle Processes and Activities"*, Version 3.2.2. San Diego, CA, USA: International Council on System Engineering (INCOSE), INCOSE-TP-2003-002-03.2.2

The Five-hundred-meter Aperture Spherical radio Telescope (FAST) project

Rendong Nan[1,2] and Di Li[1,3]

1.National Astronomical Observatories, Chinese Academy of Sciences, A20 Datun Road, Chaoyang District, Beijing 100012, China

2. Key Laboratory for Radio Astronomy, Chinese Academy of Sciences, Nanjing 210008, China

3. Space Science Institute, Boulder, CO 80301, USA

E-mails nrd@bao.ac.cn dili@nao.cas.cn

Abstract: Five-hundred-meter Aperture Spherical radio Telescope (FAST) is a Chinese "mega-science" project to build the largest single dish radio telescope in the world. Its engineering concept and design pave a new road to realize a huge single dish in an effective way. Being the most sensitive single dish radio telescope, FAST will enable astronomers to jump-start many science goals, such as surveying the neutral hydrogen in the Milky Way and other galaxies, detecting faint pulsars, hearing the possible signals from other civilizations, etc. The feasibility studies for FAST have been carried out for 14 years, supported by Chinese and international astronomy communities. The National Development and Reform Commission approved the funding proposal of FAST in 2007 with a capital budget close to 700 million RMB. The project time is 5.5 years from the commencement of work in March of 2011 and the first light is expected in 2016.

1. Engineering concept of FAST

FAST is an Arecibo-type antenna with three outstanding aspects: the karst depression used as the site, which is large enough to host the 500-meter telescope and deep enough to allow a zenith angle of 40 degrees; the active main reflector correcting for spherical aberration on the ground to achieve a full polarization; and a wide band without involving complex feed systems. If we adopt Arecibo design, we would need 10000 tons of steel above the reflector, which is implausible. We have very light focal cabin weighting less than 50 tons. Feed cabin is supported and driven by cables and servomechanism. Inside the cabin, secondary adjustable system is mounted to compensate the offset of the cabin mainly due to wind impact.

The outstanding features of FAST and its optical geometry are shown in Figure 1. Main reflector is a spherical cap with the curvature of a 300m radius and opening up to 500m. During the observation, the 300 m illuminated area forms a paraboloid to the source direction. Feed moves on a virtual spherical cap half-way from the reflector to the origin of the sphere. The central part of a spherical surface is very nearly a paraboloid of revolution when a proper focal length is chosen. In the configuration shown in the figure, the peak deviation between deformed paraboloid (yellow dashed line) and the spherical cap will be minimized to 0.67 m within the 300 m illuminated aperture. For the maximum apparent motion of celestial objects, the rate of variation is also found to be very small, less than 0.7 mm/s, which enables a straightforward solution for the mechanical control. To deform the reflector, it is necessary to divide it into small elements. Each element is a small part of a spherical

IOP Conf. Series: Materials Science and Engineering **44** (2013) 012022 doi:10.1088/1757-899X/44/1/012022

surface, and curvatures of all the elements are identical to get an isotropic geometry for the reflector. The main technical specifications are listed in table 1.

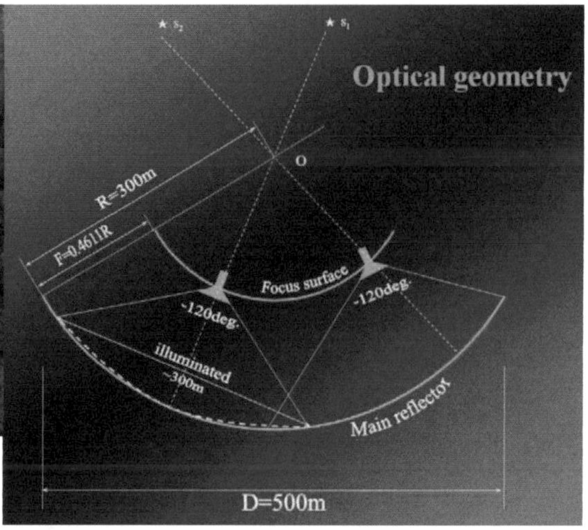

Figure 1 Left: Simulated image of FAST.
Right: The optics of the telescope.

Table1. Main technical specifications of FAST

Spherical reflector: Radius = 300m, Aperture = 500m
Illuminated aperture: D_{ill} = 300m
Focal ratio: f/D = 0.4611
Sky coverage: zenith angle 40
Frequency: 70MHz - 3GHz
Sensitivity (L-Band): A/T ~ 2000, system temperature T_{sys} ~ 20K
Resolution (L-Band): 2.9′
Multi-beam (L-Band): beam number = 19
Slewing time: <10 minutes
Pointing accuracy: 8″

2. Technical plan - critical technologies

The FAST site selection started in 1994 and a karst depression called Dawodang in the southern Guizhou Province was finally chosen.

Figure 2 Location (left), original landscape of the site (middle), current look after excavation (right).

Telescope construction is divided into six sub-systems, including site survey and excavation; active reflector system; feed cabin suspension system; measurement and control system; receiver and back-end system; and observatory buildings.

2.1. Site survey and excavation

Despite of the excellent fit between the depression profile and the spherical cap model, we still need to move some one million cube meters of earth from the karst hole. A tunnel to link the sinkhole to lower depression is required in order to guarantee the safety of the telescope. As of September 2012, the site excavation and protection of the dangerous rocks and slopes are near completion. The construction of underground tunnel of ~ 1.2 km and external roads of ~4 km in the depression have been successfully completed.

2.2. Active reflector system

Most expensive part of the telescope is the main active reflector of 500 m in diameter. The whole surface is to be supported by cable network. More than 2000 actuators drive tie-down cables according to the feedback from the measuring system to deform the surface. The optimization of the element size is a trade off between accuracy and cost. The whole reflector consists of ~ 4400 triangular panels, which give an RMS error smaller than ~ 5.0 mm.

From the aforementioned description, it can be easily seen that the long-term observation process of FAST is equivalent to long-term frequent shape-changing operation. Early research results have shown that such shape-changing operation would lead to about 500MPa of stress range, which is nearly twice the industry standard. The FAST team have carried out extensive numerical and experimental investigation to decrease the stress range of the cable required and have successfully developed a new type of steel strand with ultra high fatigue resistant property. All samples have passed through 2×10^6 fatigue cycles under 550MPa stress range. Compared with full 70000 fatigue cycles required during FAST life time and the maximum demanded stress range of 450 Mpa, enough reserve has been guaranteed (Jiang *et al.* 2011).

The side dimension of each reflector panel is about 11m, depending on its location on the surface. Different kind of prototyping have been investigated and experimented. The final design has a surface made of the aluminum sheet, backup of aluminum spatial truss and an adjustable layer in between.

It is extremely important to have reliable actuators considering their large amount. Besides long life, easy maintainability and less dependence on neighboring ones are crucial factors. Various types of actuators have been manufactured and tested at the Miyun station of NAOC and at the depression site. The final selection between mechanical and hydraulic actuators is under way.

2.3. Feed cabin suspension system

Feed cabin suspension system might be the most risky part of FAST. There is no solid connection between the reflector and the feed cabin because of the large dimension of the telescope. The feed cabin of FAST is supported and driven by cables and servomechanism. Inside of the cabin, secondary adjustable system is employed to achieve the required accuracy. The designs consist of three essential parts, including the cable network that supports and drives the feed cabin, the secondary adjustable devices inside the cabin that carry the most precise part of the receivers, and the close loop control. Numerous models were made to verify the feasibility of the concept. Because of the large dimension and the complicated dynamics, similarity laws may not guarantee applying the results from down-scaled models to real structure. The team has carried out end-to-end simulation cooperated with MT Mechatronics and Technical University Darmstadt. This analysis has confirmed the feasibility of the concept and yielded critical results for system optimization. The displacement of feed cabin after first adjustment control can be constrained to be within a few centimeters. The achievable accuracy of feed position with the help of the secondary stabilizer turns out to be a few millimeters, which meets the requirement of telescope pointing.

The detailed design and manufacture have started in a Chinese lift firm. The detailed design for 6 suspension towers higher than a hundred meters has been finished and the groundwork has begun. Preliminary design for feed cabin has completed, and detailed design and construction started to accept bids.

2.4. Measurement and control system

Major parts of FAST are active, that requires fast and high accuracy measurement and control in long distance. One heavy task is the control of 3-D positions and orientation of focus cabin which need high accuracy ~ 2 mm, sampling rate > 10 Hz in a large working range up to 300 m. Another one is the control of the profile of main reflector, surveying ~1000 nodes in illuminated area in real time within a interval of a few minutes.

The datum lines have been established, monitoring shows the accuracy and stability better than 1 mm. Complex control network underneath the reflector is at the concept design stage, dividing the ~2400 actuators into districts and groups. Most challenging issue would be the EMI from active equipments over the depression.

2.5. Receiver and back-end system

FAST will build eight sets of receivers, covering a frequency range of 70MHz-3GHz. Scientific back-ends, time and frequency standard, monitoring/diagnostics of the receivers have been optimized through long term international cooperation. The 19-multi-beam receiver at L-band is the core receiver system of FAST. A trilateral collaboration among NAOC, JBCA and CSIRO has been established for a designing study of this important receiver.

L-band single pixel receiver has been developed in FAST lab, covering a wide band 1.1 – 1.9 GHz with return loss better than -22dB and RL isolation < -22dB across the band. We are now working on the feasibility of using honeycomb and carbon fibre material to reduce the weight, therefore reduce the load on the stabilizer in the cabin.

3. FAST Sciences

FAST should have significant impact on astronomy and has the potential to revolutionize other areas of the natural sciences. Its unique contributions to science may not yet be predictable at present. Major key scientific projects are planned (Nan *et al.* 2011)

- survey the Galactic ISM in HI at a resolution comparable to the current large scale CO surveys;
- discover ~ 4000 new Galactic pulsars and search for the first extragalactic pulsars beyond Magellanic clouds;
- detect hundreds of thousands of HI galaxies and detect individual massive galaxies up to z ~ 1;
- spectroscopic survey of rich Galactic sources with continuous coverage between 70MHz and 3GHz;
- join the international VLBI experiments
- search for radio signals from exoplanets
- SETI

During the early stages of telescope adjustments, we could start from low frequencies. Two receivers are planned for early sciences. One is the 7- multi- beam receiver with an operating band 400MHz-560MHz; and another single pixel 270MHz-1450MHz. These receivers will be used for low frequency deep pulsar survey in the Milkyway and nearby galaxies, for searching for large molecules in Orion nebula, for studying HI distribution of nearby disk galaxies, e.g. M31 and M33, and for detecting distant OH mega-masers.

Acknowledgement

This work is supported by China Ministry of Science and Technology under State Key Development Program for Basic Research (2012CB821800).

References

[1] Jiang P 2011 Applied Mechanics and Materials **94** 979
[2] Nan R D, Li D, Jin C J, Wang Q M, Zhu L C, Zhu W B, Zhang H Y, Yue Y L & Qian L 2011 International Journal of Modern Physics D **20** 989

IOP Conf. Series: Materials Science and Engineering **44** (2013) 012023 doi:10.1088/1757-899X/44/1/012023

Wideband feeds for the upgraded GMRT

Hanumanth Rao Bandari, G. Sankarasubramanian and A. Praveen Kumar

GMRT Observatory, NCRA-TIFR, Post Bag No. 3, Ganeshkhind, Pune – 411007, Maharashtra, India

Emails: hanuman, shankar, praveen@gmrt.ncra.tifr.res.in

Abstract: This paper describes the existing feeds in use at the GMRT Observatory and details the ongoing development of next generation wideband feeds for the upgraded GMRT. The existing feeds include: feed with folded thick dipoles (for 150 MHz), dipole-disc feed (for 325 MHz), dual-band coaxial feed (for 233 MHZ and 610 MHz), and corrugated horn feed (for 1400-1450 MHz). The new broadband feeds covered in this paper are: cone-dipole feeds for 250-500 and 500-1000 MHz, wideband horn feed for 550-900 MHz, and dual ring feed for 130-260 MHz. Design techniques and performance results for these are described.

1. Introduction

The Giant Metrewave Radio Telescope (GMRT) is an international facility for Radio Astronomy, operational since 2002. Located near Pune in western India, the GMRT consists of 30 fully steerable parabolic dishes, each of 45 m diameter, spread over an area with an effective radius of about 15 km. It operates in 5 distinct bands covering frequencies in the range of 50 MHz to 1420 MHz [1]. Though meant primarily as an aperture synthesis instrument, antenna arrays can be formed out of the 30 dishes, where separate sub-arrays can work on different frequency bands concurrently.

The GMRT has prime-focus feeds at four faces of a rotating turret, covering frequency bands at 150, 233, 327, 610 & 1420 MHz (one turret face houses a dual frequency feed operating at 233 & 610 MHz), with relatively modest bandwidths of a few tens of MHz, except for the 1420 feed that covers 1000 to 1450 MHz [2]. Currently, the GMRT is going through an upgrade to provide seamless frequency coverage from 50 to 1500 MHz, with a maximum instantaneous bandwidth of 400 MHz for increased sensitivity. For this, new broadband feeds are being designed with octave bandwidths.

2. The existing narrowband feeds at GMRT

The existingfeeds in use at the GMRT are as follows:

(a) The 150 MHz band feed consists of a pair of thick folded dipoles over a ground plane (figure 1). It gives good VSWR over the frequency band 130-230 MHz, with an edgetaper of -9dB, and a cross-polar of -17 dB. However, good match of E& H plane patterns and acceptable beamshape are obtained only over a narrower range of130-150 MHz.

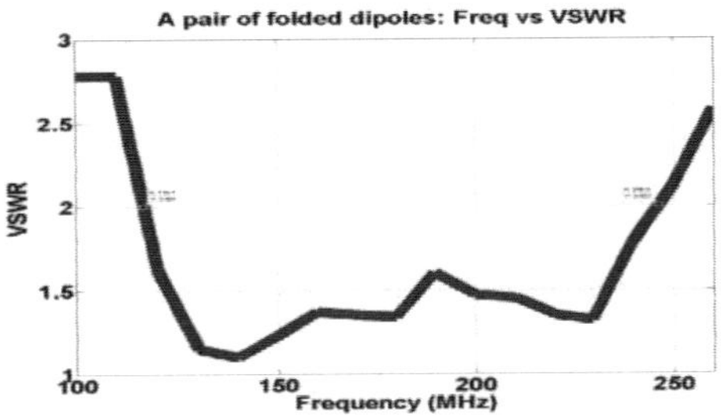

Figure 1: 150 MHz feed (a) feed on antenna (b) measured VSWR response

(b) The 235/610 MHz band feed is a dual coaxial waveguide feed (figure 2). It covers the frequency bands 225-245 MHz and 580-780 MHz, with an edgetaper of -9.8 dB and a cross-polar of -22.8 dB.

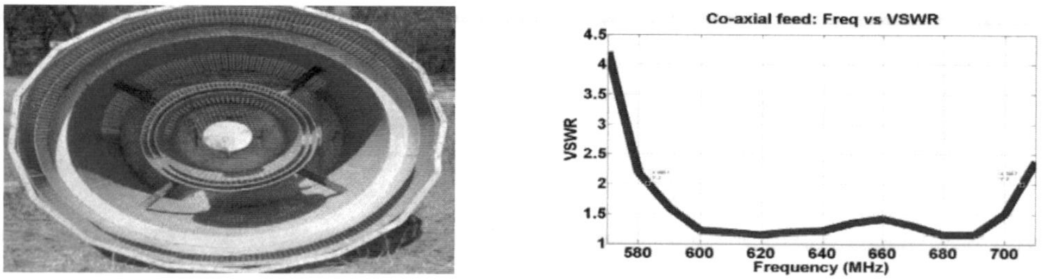

Figure 2: 235/610 feed (a) photograph of feed (b) measured VSWR response

(c) The 325 MHz band feed is a dipole-disk antenna with a beam-forming ring[3] (figure 3). It covers the frequency band 290-420 MHz, with an edgetaper of -12.2 dB and a cross-polar of -27.5 dB.

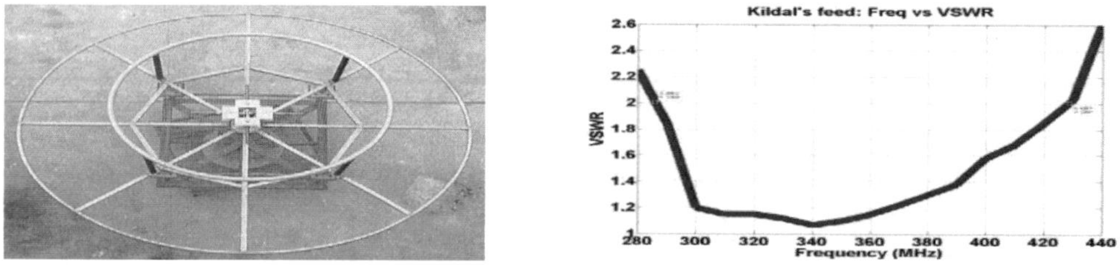

Figure 3: 325 feed (a) photograph of feed (b) measured VSWR response

(d) The 1420 MHz band feed is a corrugated horn with quad-ridged orthomode transducer (figure 4).This feed covers 1000-1450 MHz with an edgetaper of -19 dB and a cross-polar of -24 dB.

Figure 4: 1420 feed (a) mounted on the antenna (b) detailed view of the feed

3. New broadband feeds for the upgraded GMRT

As part of the ongoing upgrade of the GMRT, the plan is to provide seamless frequency coverage from 50 MHz to 1500MHz.To cover this range efficiently, the following wideband feeds (octave or more bandwidth) are being designed (using the WiPL-D software [4]): 130-260 MHz, 250-500 MHz and 500-1000 MHz. The details of these are as follows.

(a) For 250-500 MHz, the prototype feed that has been designed and tested employs a crossed-dipole with a conical reflector [5] (figure 5). Test results indicate a BW ratio of 1:1.8 for a return-loss of -10 dB or less (figure 6), with the following radiation properties: (1) good pattern symmetry and fairly good match of E and H plane patterns (on the primary lobe, especially within the angular spread of -62.5 to + 62.5 deg., being the edge-angle of the GMRT's parabolic dish) over 250 to 500 MHz(2) a cross-polar minimum of -27 dB at 327 MHz and -15.5 dB at 250 as well as 500 MHz. After extensive tests of prototypes on 3 GMRT antennas (figure 7), this feed is now accepted for mass production.

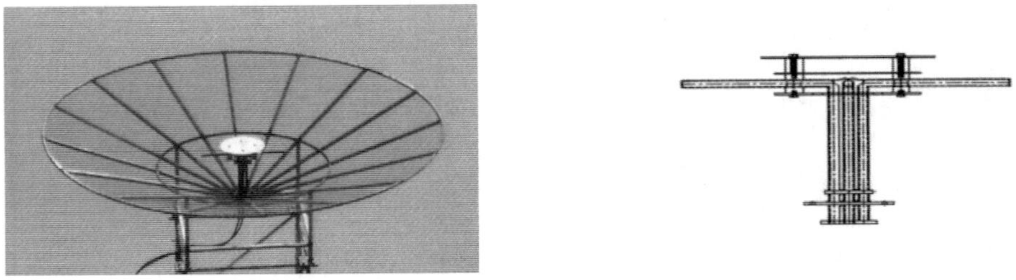

Figure 5: 250-500 feed (a) mounted on antenna (b) Schematic of dipole with triple-sleeve

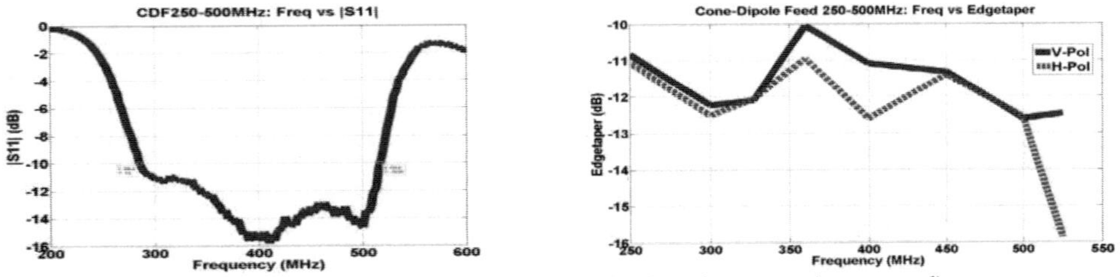

Figure 6: 250-500 feed (a) Return Loss (measured) (b) edgetaper (measured)

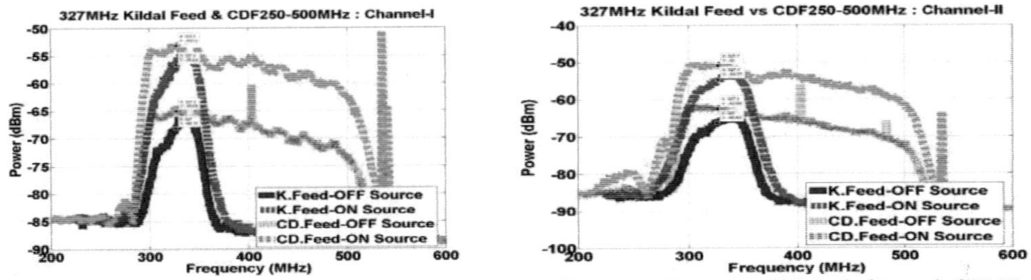

Figure 7: 250-500 feed: Deflection tests on calibrator (Cass-A) CH-I (left) and CH-II (right); the narrower curves are for the existing narrow band 325 MHz feed.

(b) For 500-1000 MHz, a scaled version of the 250-500 MHz feed with crossed-dipole and conical reflector is being tried out as a possible option. The present version (figure 8) covers 580-1000 MHz. Some more optimization of the performance is needed before it can be accepted for final use.

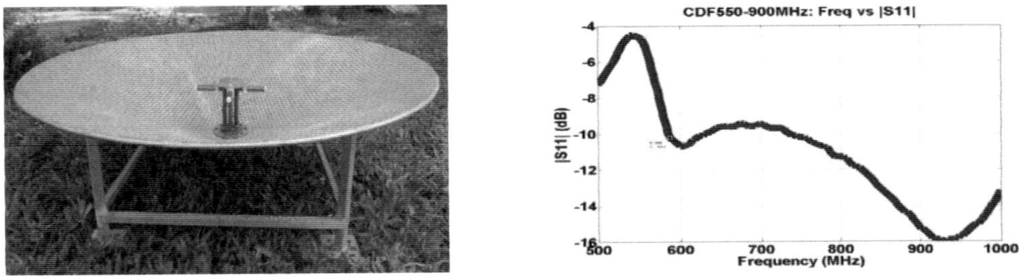

Figure 8: 500-1000 feed (a) photograph of feed (b) measured Return Loss response

(c) A horn feed alternative for 500-1000 MHz, designed in collaboration with CSIRO, Australia, has been built and is being tested (figure 9). It covers 550-900 MHz with an edge taper of -12 dB.

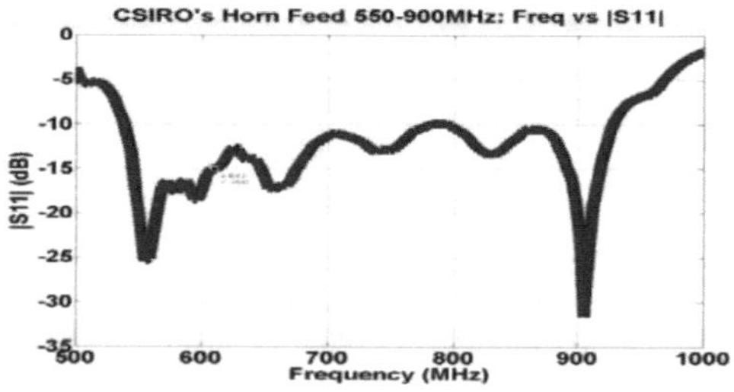

Figure 9: 550-900 feed (a) feed on antenna (b) measured Return Loss response

(d) For 130-260 MHz coverage, the basic design of the Kildal Feed (dipole-disk feed) is being adopted, with the following design changes: (i) sleeved cross dipoles for larger BW (ii) provision of 2 beam-forming rings for better polarization features over a larger BW and (iii) an optimized reflector design.The prototype version of this feed (figure 10) covers 138-279 MHz, and is quite promising.

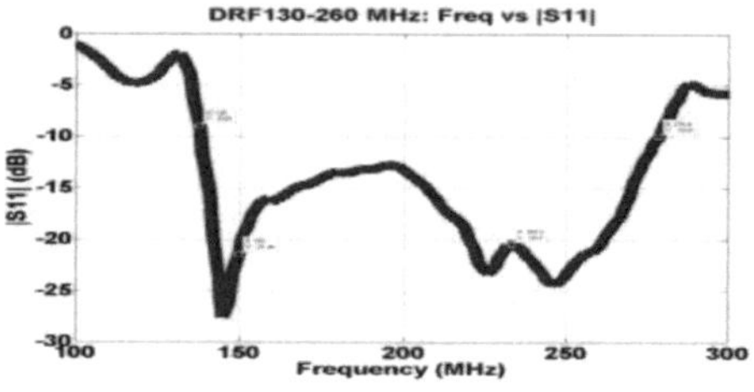

Figure 10: 130-260 feed (a) photo of feed (b) measured Return Loss response

4. Summary and Conclusions

Existing narrowband feeds at the GMRT provide 5 observing bands in 130 to 1500 MHz range and are working well for the last 15 years.As part of the ongoing upgrade of the GMRT, to provide seamless frequency coverage from 50 to 1500 MHz; new feeds are being developed to cover the following bands: 130-260 MHz, 250-500 MHz and 500-1000 MHz. The cone-dipole feed designed for 250-500 MHz is found to work well and is in mass production. A similar feed is being tried out for 500-1000 MHz, along with a horn feed (550-900 MHz); and a dual-ring feed design for 130-260 MHz is promising.Successful completion and deployment of these wideband feeds on all 30 antennas will make the GMRT a very sensitive and versatile instrument for a variety of new science in astronomy.

5. Acknowledgements

We thank S. Suresh kumar and colleagues at the front-end lab of the GMRT for their support, WiPL-D team for technical support, workshops at NCRA & GMRT for fabrication and assembly of the prototype feeds, and Prof. Y. Gupta for comment on this paper. The GMRT is operated by the NCRA.

References

[1] Swarup G, et al. 1991 "The Giant Metre-wave Radio Telescope," *Current Science,* **60**, 2
[2] Chengalur J N, Gupta Y and Dwarkanath K S 2007 Low Frequency Radio Astronomy 3rd Edition NCRA-TIFR Pune
[3] Kildal P-S and Skyttemyr S K, 1982 "Dipole-Disk Antenna with Beam-Forming Ring," *IEEE Trans. Ant. and Propg.,* **AP-30**,4, pp.529-535
[4] Kolundzija B, Ognjanovic J, Tasic M, Olcan D, Sumic D, Paramentic M, Kostic M, Pavlovic M, "WIPL-D Pro v7.0, 3D Electromagnetic Solver Prof. Edition," WIPL-D Ltd. 11070, Belgrade
[5] WongJ L and King H E, 1973, "A Cavity-Backed Dipole Antenna with Wide-Bandwidth Characteristics," *IEEE Trans. Ant. and Propg.*, **AP-21**, 5, pp. 725-727

IOP Conf. Series: Materials Science and Engineering **44** (2013) 012024 doi:10.1088/1757-899X/44/1/012024

Next Generation Digital Backends for the GMRT

B. Ajithkumar, S.C. Choudhari, K.D. Buch, M.V. Muley, G.J. Shelton, S.H. Reddy, S. Kudale, J.Roy and Y. Gupta

GMRT Observatory, NCRA-TIFR, Post Bag No 3, Ganeshkhind, Pune-411007, Maharashtra, INDIA

Email : ajit@ncra.tifr.res.in

Abstract : The GMRT Backend Group is in the process of developing wide-band digital backends as part of a major upgrade of the GMRT. The main requirements are for processing 400 MHz bandwidth signals from 30 antennas for interferometry and array mode operations, along with additional features like narrow band modes and RFI cancellation schemes. Both FPGA and GPU based designs have been explored and this paper presents the design and implementation details along with results from testing of prototype units. A comparison of the resources needed and performance achieved on FPGA and GPU based designs is presented, along with an extrapolation of these designs to a 30 station backend for the final GMRT system.

1. Introduction

The Giant Metrewave Radio Telescope (GMRT) is a large, fully steerable telescope array of thirty 45 m diameter antennas, built for studying the astrophysical phenomena at low radio frequencies [1]. Located near Khodad, at a distance of about 80 kms north of Pune city in India, the GMRT array spread over a 30 km region, consists of a compact central area of 1 km by 1 km having 12 dishes, and the remaining 18 are distributed in 3 arms of a "Y" shaped array, that extends 15 km from the centre. Astronomical observations are currently possible at 5 different RF bands – 150, 235, 325, 610, 1420 MHz – with an instantaneous bandwidth of 32 MHz from each of two polarisation channels. The present digital backend that processes these signals to implement a 30 antenna, full polar correlator (for interferometry requirements) and incoherent / coherent beamformer (for array mode requirements) is the GMRT software backend (GSB) that utilizes a network of CPUs to carry out the tasks in real-time, achieving a peak performance of about 400 GFlops [2]. The GMRT observatory is undergoing a major upgrade to provide seamless frequency coverage for observations between 50 to 1500 MHz with instantaneous bandwidth of 400 MHz. New digital backends are being developed for processing the signals from antennas with 400 MHz bandwidth to produce interferometry and array outputs. The compute requirement for this is estimated to be about 6 TFlops.

2. Digital Backend – Implementation Options

The main specifications for the new backend system being developed for the upgraded GMRT are as follows : 30 inputs, dual polarisation, 400 MHz processing bandwidth, 8-bit sampling, up to 8192 spectral channels, delay correction range of +/- 128 microsec, fringe correction up to 5 Hz, correlator dump times 1 sec or better, full Stokes capability, array mode with incoherent and coherent beamformer options for pulsar processing, narrow band modes and RFI excision capabilities, raw voltage recording and playback options. Two possible options for implementing the system have been tried out and are described in this paper.

2.1. Field Programmable Logic Array (FPGA) based design
This design is based on the packetised design approach of CASPER (Centre for Radio Astronomy Signal Processing and Electronic Research) of which the GMRT backend team is a member [3]. An 8-antenna, dual polarisation FPGA based digital correlator has been implemented and tested at GMRT in collaboration with the SKA-SA team from South Africa.

2.2. Graphical Processing Unit (GPU) based design
This is a hybrid design where the FPGA does the data capture and sends it on a 10GbE link to a host computer with a high performance GPU that does the bulk of the compute intensive data processing. An 8-antenna dual polarisation design of such a correlator has been developed in collaboration with Swinburne University of Technology, Australia and nVidia India, and has been tested at the GMRT.

Both options use the hardware boards from the CASPER collaboration [3] : iADC which uses Atmel make dual 8-bit ADC chips and a generic FPGA board called ROACH (Reconfigurable Open Architecture Computing Hardware) which uses one Virtex-5 chip for signal processing.

3. FPGA based design
The basic block schematic of the FPGA implementation of the digital correlator is shown in figure 1. The main components are dual channel ADCs for digitising both polarisation channels from each antenna, and ROACH boards for the processing – one set called F-engines which do the coarse delay correction, FFT, fine delay (FSTC) and fringe corrections, and the second set called X-engines which carry out the multiply and accumulate (MAC) operations and send the outputs to a host computer for storage. A 10 GbE switch is used to transfer the data from the F-engines to the X-engines such that each X-engine processes a pre-defined slice of the complete frequency spectrum from all the antennas.

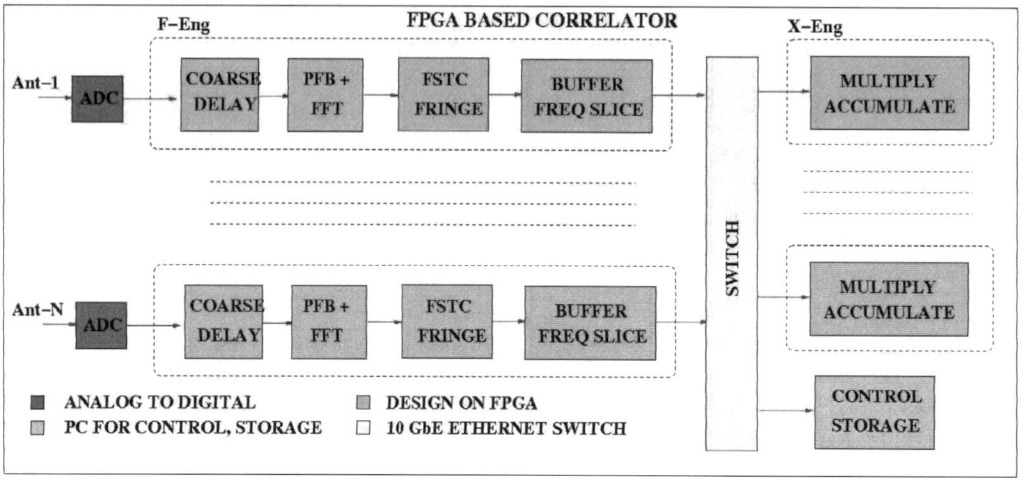

Figure 1 : Schematic block diagram for a FPGA based backend

An 8-antenna dual polarisation correlator based on this FPGA design has been implemented using 8 ADC boards, 8 F-engine and 8 X-engine ROACH boards, and tested with GMRT antenna signals. A typical self and cross spectrum of 110 MHz bandwidth signal at L-band from two antennas (E02, E06 in the eastern arm of the GMRT) while tracking the astronomical source 3C286, with an accumulation time of 2 sec, is shown in figure 2. Detailed tests and comparison with the existing 32 MHz GMRT Software Backend, show matching results (figure 3) which validates the basic performance of this new broadband backend design. By extrapolating this design, a version meeting the specs for the 30 antenna GMRT has been outlined. The estimated cost of this design is USD 469,400 and the net power dissipation is estimated to be about 5830 W.

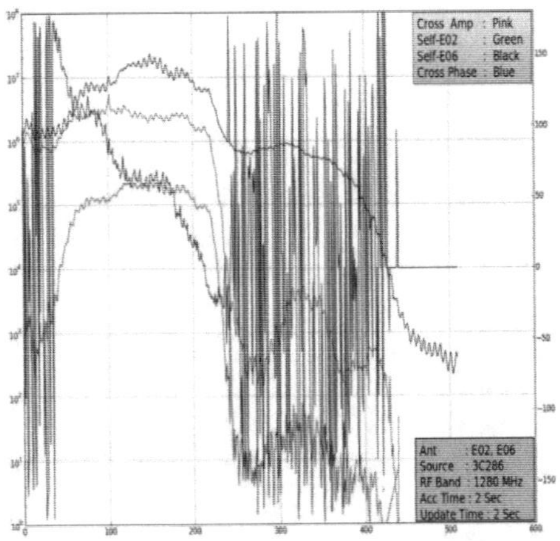

Figure 2 : Spectrum of FPGA Correlator

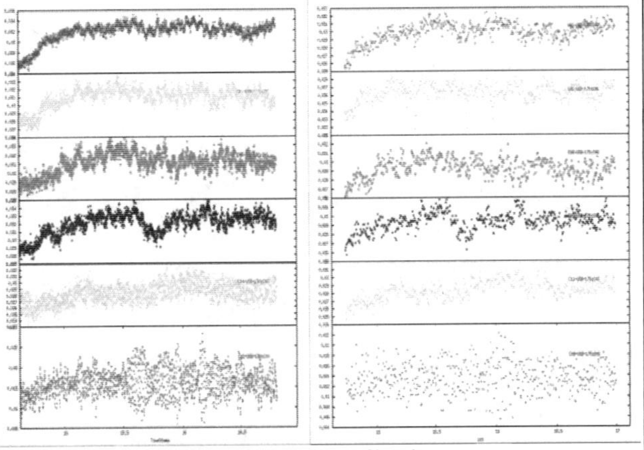

Figure-3 : Stability of cross amplitude : FPGA vs Software Backend at 1280 MHz on 3C286

4. GPU based design

GPUs consist of thousands of small efficient cores optimised for parallel performance. A CPU+GPU node is a powerful combination where the compute intense portion of the code is offloaded to GPU for processing while the CPU host handles the remaining portions, including i/o operations. At the GMRT we have implemented a GPU based 8-antenna dual polarisation correlator and tested it using GMRT signals. In our implementation we have used C2050 GPU cards of nVidia [4] with Dell T7500 machines as host. The ADC + FPGA board combination is used for digitizing and packetizing the input voltage signals from 2 channels, which are then sent to one host PC via a 10GbE connection. Such data from 8 host PCs are then shared over a 10GbE switch to provide each PC with a small time-slice of the full data block, from all the antennas (figure 4). The time slices are transferred to the GPU, where all the F-Engine and X-engine processing happens, and the final integrated results are passed on to the host and from there to a single machine for storage after stitching together the different time slices. For efficient handling of the tasks, the host machine employs multiple threads and MPI synchronization.

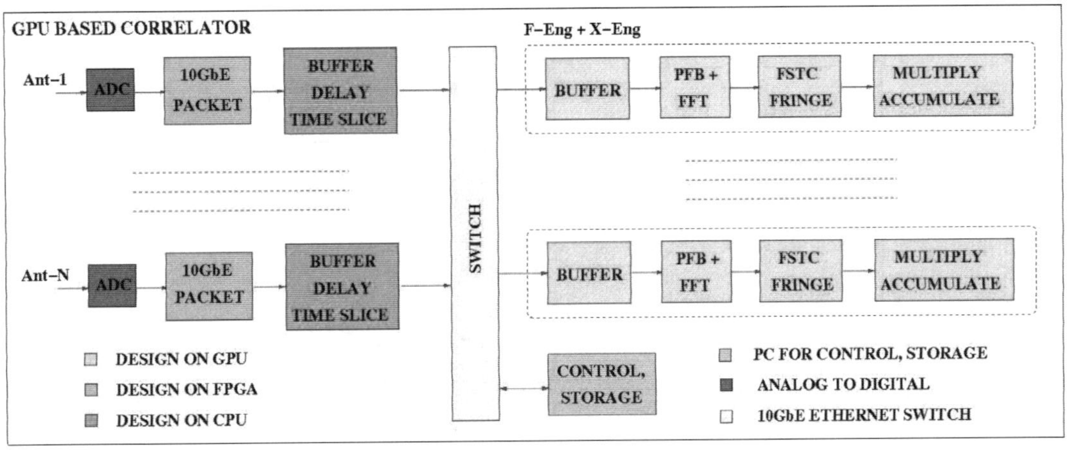

Figure-4 : Schematic block diagram for a GPU based backend

Test data acquired with the GPU correlator has been used to generate radio maps of sources in the sky and these have been found to compare favorably with maps from GSB data (figure 5). By extrapolation, the estimated cost of this design for a 30 antenna system is expected to be USD 491600 and power consumption is expected to be 20,800 W.

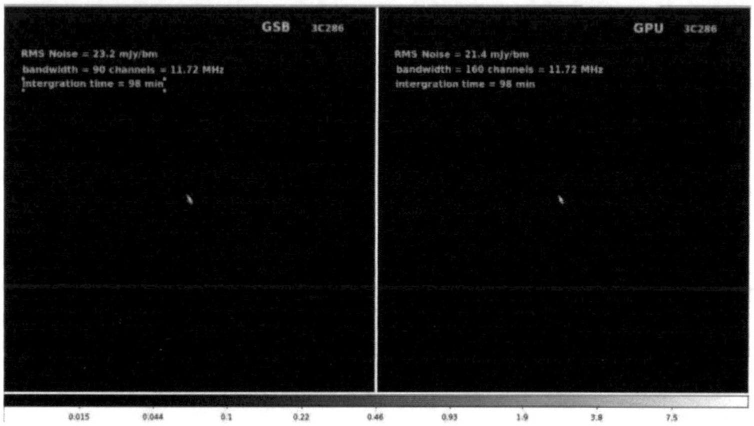

Figure-5 : Comparison - Image from GSB & GPU

Source : 3C286
8 Ant, 1 pol used
RF : 1280Mhz

Source size, Location, RMS noise, Flux level match between GSB and GPU images.

5. Summary and Conclusion

The GMRT team has successfully developed 2 kinds of prototype designs for 400 MHz bandwidth backend systems. Based on our experience with these FPGA and GPU based correlator prototypes, it can be concluded that for a full GMRT backend, cost implications are similar in both FPGA and GPU approaches, but the total power consumption in GPU based designs is expected to be about 4 times higher. At the same time GPU designs are more flexible and the parameter set can be changed and new features added easily without major redesign effort, and they also benefit from full floating point arithmetic.

Acknowledgements

We thank Francois Kapp, Jason Manley and their team members from SKA-SA in South Africa for work on the FPGA design; Ben Barsdell and Andrew Jameson from the team of Matthew Bailes at Swinburne University of Technology, Australia who have been involved in the GPU design; and colleagues from nVidia India who have helped in optimization of the GPU code and have provided resources for development work on GPUs. The GMRT is run by the National Centre for Radio Astrophysics of the Tata Institute of Fundamental Research.

References

[1] Swarup G, et al, 1991The Giant Meterwave Radio Telescope *Current Science* **60** No.2 pp 95-105
[2] Roy J, Gupta Y, Pen U L, Peterson J B, Kudale S and Kodilkar J. 2010 *Experimental Astronomy* **28** pp 25-60
[3] Casper website : https://casper.berkeley.edu/wiki/CASPER_Correlator
[4] nVidia website : http://www.nvidia.in/page/home.html

Radio and Antenna Days of the Indian Ocean (RADIO 2012) IOP Publishing
IOP Conf. Series: Materials Science and Engineering **44** (2013) 012025 doi:10.1088/1757-899X/44/1/012025

Front-end Electronics for the upgraded GMRT

Anil N. Raut, Vilas Bhalerao, A. Praveen Kumar

GMRT Observatory, NCRA-TIFR, Post Bag No. 3, Ganeshkhind, Pune – 411007, Maharashtra, India

Emails: anil, vilas, praveen@gmrt.ncra.tifr.res.in

Abstract : This paper first describes briefly the existing front-end receiver in use at the GMRT observatory and then details the ongoing development of next generation receiver systems for the upgraded GMRT. It covers the design of the new, two stage, room temperature, low noise amplifiers with better noise performance and matching, and improved dynamic range that are being implemented for the 130-260 MHz, 250-500 MHz and 550-900 MHz bands of the upgraded GMRT front-end systems.

1. Introduction

The GMRT is an aperture synthesis array of 30 fully steerable, 45 metre diameter antennas, spread out over a 30 km region near Pune in western India [1]. It is designed to operate at a range of frequencies from 50 MHz to 1420 MHz, in five distinct frequency bands. All the front end electronics for these bands is located along with the feeds at the prime focus of the dishes, on a rotating turret [2]. Currently the GMRT is going through an upgrade to provide seamless frequency coverage from 50 to 1500 MHz, with a maximum instantaneous bandwidth of 400 MHz for increasing the sensitivity of the telescope. As a part of this upgrade, the existing front end receivers are being modified to have wide band low noise amplifiers (LNAs), octave-band polarizers with low insertion loss, wide band filters and high dynamic range broadband post-amplifiers. The L-band front end receiver has already been reconfigured and modified for better dynamic range, using a high dynamic range two stage wideband post amplifier, and band reject filter for rejecting CDMA and GSM mobile interference signals. New LNAs are being designed for the other broadband systems that are part of the upgraded GMRT.

2. The existing GMRT front end receiver system

Existing GMRT front-end systems operate at five frequency bands centered at 150 MHz, 233 MHz, 327 MHz, 610 MHz and L-band extending from 1000 to 1420 MHz [2]. The L-band is split into four

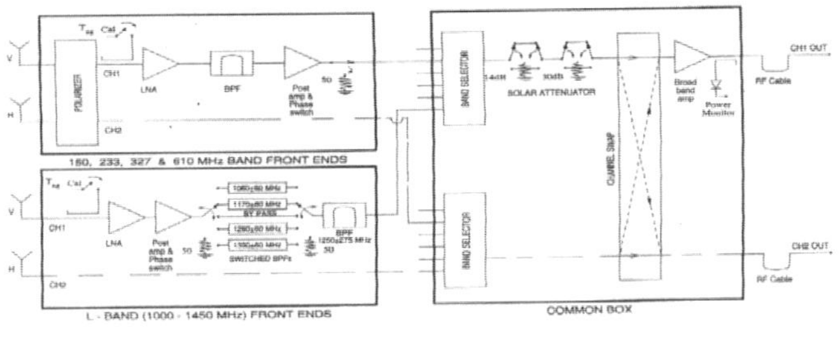

Figure 1. Block diagram of existing GMRT Front End Receiver

sub-bands centered at 1060 MHz, 1170 MHz, 1280 MHz and 1390 MHz, each with a bandwidth of 120 MHz (see figure 1). The 150 MHz, 233 MHz and 327 MHz bands have about 40 MHz bandwidth, and the 610 MHz band has about 60 MHz bandwidth. All these low noise front-end receivers operate at ambient temperature. They are designed for dual polarization signals, with the lower frequency bands from 150 to 610 MHz having dual circular polarization channels and the higher frequency L-band having dual linear polarization channels. The front end system has flexibility to be configured for either dual polarization observation at single frequency band or a single polarization observation at two different frequency bands. The polarization channels can be swapped whenever required. For observing strong radio sources like the Sun, selectable solar attenuators of 14 dB, 30dB or 44 dB are provided. The front-end also has a RF termination facility. Any band of the receiver can be switched OFF whenever not in use, with the RF ON/OFF facility provided in the front end. The receiver can be calibrated by injecting four levels of calibrated noise, named Low cal, Medium cal, High cal and Extra-high cal, depending on the specific needs. To minimize cross coupling between channels, a phase switching facility using Walsh function modulation has been provided.

3. Front End Receiver for the upgraded GMRT

As a part of the ongoing upgrade to wideband the GMRT observatory, the front end receiver system is being modified to include wide band, high dynamic range LNAs, wide band filters and octave band polarizers with low insertion loss, amongst other improvements.

At present, new LNAs have been designed for the 130-260 MHz, 250-500 MHz and 550-900 MHz bands of the upgraded GMRT. For radio astronomy requirements, the LNA should be able to amplify extremely low level signals with minimal noise contribution, should have good input impedance match and must be unconditionally stable. Though cryogenically cooled LNAs provide the best performance, they are a costly solution and also increase the complexity of the front end receiver. For metre wavelength radio telescopes like the GMRT, where sky and background noise contributions are substantial, it is possible to obtain optimal system noise temperature performance using the latest devices at room temperatures. Requirements for higher dynamic range, characterized by high OIP3 values which are attained by higher currents in the amplifier leading to elevated noise figure, also need to be balanced with low noise temperature requirements [3].

It is found that E-pHEMT technology provides a combination of high gain, low noise and high dynamic range in high linearity LNA applications [4]. Therefore all these low noise amplifiers are designed using Avago Technologies make E-pHEMT, ATF 54143 device. The designs are simulated using Applied Microwave Research (AWR) make Microwave Office software. Two stage amplifiers are designed with single ended +5 V supply and current consumption of less than 120 mA. All these LNAs have been tested thoroughly on the bench as well as on GMRT antennas and give unconditionally stable performance. They are all in the phase of mass production at present.

3.1. 130-630 MHz Low Noise Amplifier

This LNA is designed for the upgraded 130-260 MHz front end, using some of the ideas given in [5].

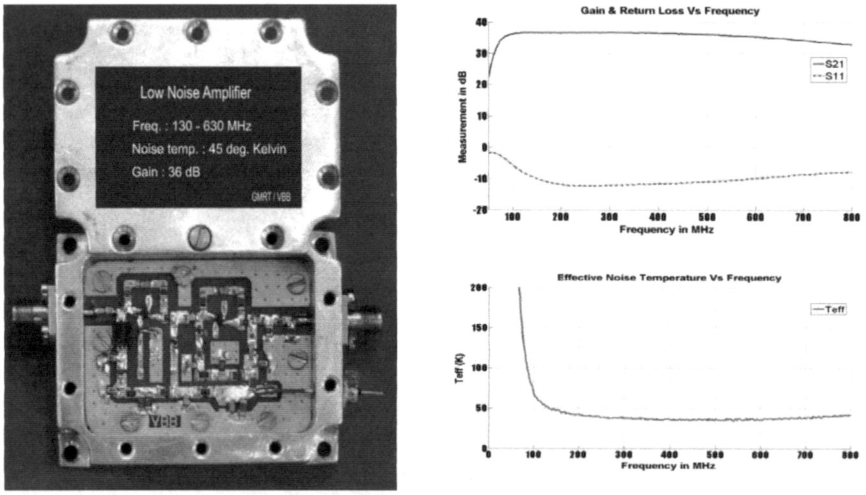

Figure 2. 130-630 MHz Low Noise Amplifier with performance curves

The LNA has a measured noise temperature (Teff) of 45 ˚K, Gain (G) of 37 dB and return loss of -10 dB (figure 2). These LNAs are integrated with the feed and rest of the receiver electronics and successfully tested on GMRT antennas. Inclusion of this LNA in the system has reduced the system noise temperature (Tsys) to 407 ˚K for 130 MHz bandwidth at ambient temperature (see table 1).

3.2. 200 – 500 MHz Low Noise Amplifier

This LNA is designed for the upgraded 250-500 MHz front end receiver, using ideas described in [5]. It has a measured noise temperature (Teff) of 20 ˚K, Gain (G) of 35 dB and return loss of -10 dB (see Figure 3). The measured OIP3 of this amplifier was found to be +18dBm. These LNAs have

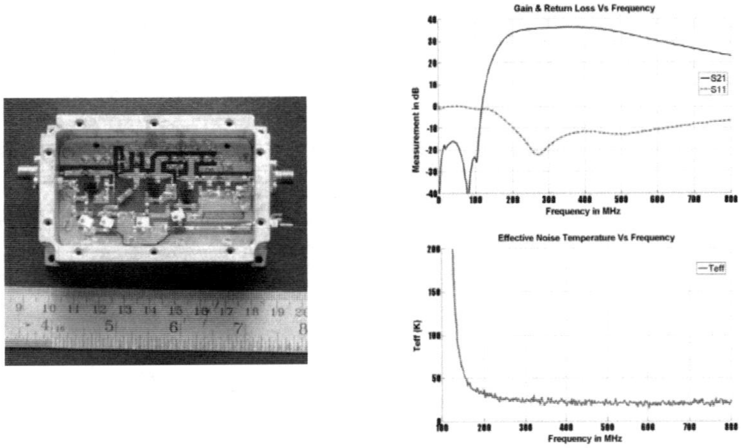

Figure 3. 200-500 MHz Low Noise Amplifier with performance curves

been integrated with the 250-500 MHz Cone Dipole Feed and rest of the receiver electronics and tested successfully on few GMRT antennas. Inclusion of this LNA in the system has reduced the system noise temperature (Tsys) to 88 ˚K for 250 MHz bandwidth at ambient temperature (see table 1).

3.3. 550 – 900 MHz Low Noise Amplifier

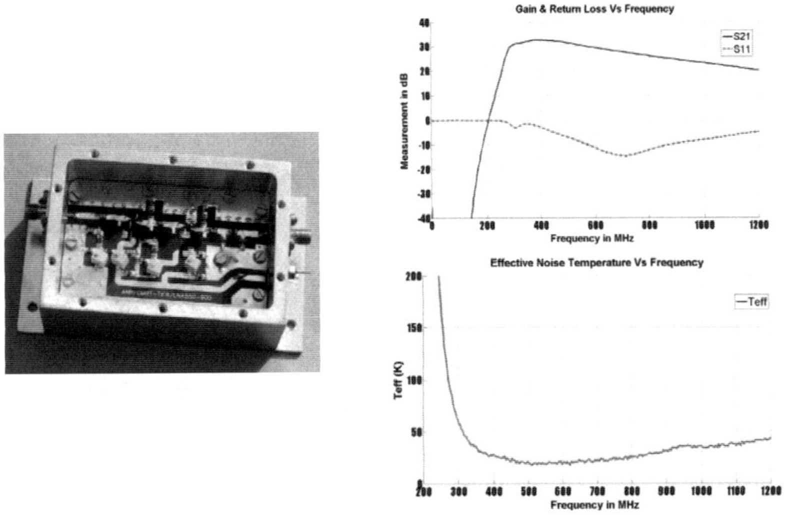

Figure 4. 550-900 MHz Low Noise Amplifier with performance curves

This LNA is designed for the upgraded 550-900 MHz front end receiver, using some of the ideas in [6]. It has measured noise temperature (Teff) of 20 °K, Gain (G) of 30 dB and return loss of -10 dB. The measured OIP3 of the amplifier was found to be +18dBm. These LNAs are integrated with 550-900 MHz Circular Waveguide Feed designed by CSIRO, Australia and rest of the receiver electronics and tested successfully on GMRT antennas. Inclusion of this LNA in the system has reduced the system noise temperature (Tsys) to 58 °K for 350 MHz bandwidth at ambient temperature (table 1).

Table 1. Tsys values for various frequency bands of existing and upgraded GMRT

Sr. No	Frequency Band [MHz]	Input cable loss L^1[dB]	Polarizer Loss L[dB]	LNA Temp T_{LNA} [K]	Receiver Temp.[a] T_R[K]	Ground Temp. T_{Gnd}[K]	Sky Temp T_{Sky} [K]	System Temp. T_{Sys}[K]	Band width [MHz]	System Temp. (Existing FES) T_{Sys}[K]
1	150	0.1	0.4	45	87	12	308	407	130	580 (40 MHz)
2	235	0.1	0.4	45	87	12	99	198	130	234 (40 MHz)
3	327	0.1	0.1	20	35	13	40	88	250	108 (40 MHz)
4	610	0.1	0.1	20	35	13	10	58	350	101 (40 MHz)
5	1060	0.22	-	35	53	25	5	83	120	83 (120 MHz)
6	1170	0.22	-	32	49	24	4	77	120	77 (120 MHz)
7	1280	0.22	-	30	47	23	4	74	120	74 (120 MHz)
8	1390	0.22	-	28	45	23	4	72	120	72 (120 MHz)

[a] Includes Input cable losses

4. Summary

New wideband, high dynamic range, room temperature LNAs have been successfully designed for 3 of the wideband configurations of the upgraded GMRT. They have been integrated with the feed and front-end electronics systems and have passed the initial acceptance tests and are in mass production phase. There is scope of improvement in the gain slope of 550-900 MHz LNA by incorporating a negative feedback network in the second stage, and possibility of reducing the Teff for 130-260 MHz LNA. Successful completion of these will improve the sensitivity and versatility of the GMRT by a large amount, keeping it at the forefront of low frequency radio astronomy at the global level.

5. Acknowledgements

We wish to express our gratitude to colleagues who are involved directly or indirectly in the success of the GMRT upgrade works. Special thanks to our friends in the front-end Lab and mechanical team at GMRT and NCRA for their constant support. Thanks to Prof. Yashwant Gupta, Mr. G. Shankar and Mr. S. Sureshkumar for helping us improve the presentation of this paper. The GMRT is operated by the National Centre for Radio Astrophysics.

References

[1] Swarup G, et al, 1991The Giant Meterwave Radio Telescope *Current Science* 60 No.2 pp 95-105
[2] Chengalur J N, Gupta Y and Dwarkanath K S 2007 *Low Frequency Radio Astronomy* 3rd Edition NCRA-TIFR Pune
[3] Gonzalez G 1997 *Microwave Transistor Amplifier: Analysis and Design* 2nd Edition pp 294-303
[4] Ian Piper 2004 Build An E-pHEMT Low Noise Amplifier Microwave & RF pp 92-102
[5] Application note AN-5057 A 100 MHz to 500 MHz Low Noise Feedback Amplifier using ATF-54143
[6] Application note AN-1222 A low Noise High Intercept Point Amplifier for 1390 to 1990 MHz using ATF-54243 pHEMT

Developments of next generation monitor and control systems for Radio Telescopes

J Kodilkar[1], R Uprade[1], S Nayak[1], Y Wadadekar[1], J Chengalur[1] and Y Gupta[1]
[1] National Centre for Radio Astrophysics, Tata Institute Of Fundamental Research, Post Bag 3, Ganeshkhind, Pune 411007, India.

E-mail: jitendra,rajsingh,nayak,chegnalu,ygupta@ncra.tifr.res.in

Abstract. As part of the ongoing upgrade of the GMRT observatory, the monitor and control (M&C) system is being upgraded to a modern specification driven system. The basic building block of the proposed M&C framework is a SACE node which provides command, response and event data streaming interfaces to the *child* and *parent* nodes running locally or remotely in a heterogeneous operating system environment. A prototype M&C system formed by hierarchically composing SACE nodes at different levels has been successfully tested at the GMRT. For the recently built 15m antenna at NCRA, a generic, web based M&C system has been developed which allows remote, authenticated operation. We discuss issues relevant to the development of the next generation M&C systems for radio telescopes using the lessons learned from these two systems. We also summarize flexible, reusable and cost-effective approaches using off the shelf packages and technologies used in generic frameworks, which can contribute to form the basis for M&C systems of very large radio telescopes like the SKA.

1. Introduction

The Giant Metrewave Radio Telescope (GMRT) built and operated by NCRA [1] is undergoing a major upgrade in several aspects like seamless frequency coverage from 150 to 1500 MHz, large instantaneous bandwidth of observations (maximum of 400 MHz), revamped servo system and other matching improvements in support systems. The GMRT monitor and control (M&C) system is a central supervisory system which controls and monitors telescope subsystems like analog receivers, digital signal processing back-ends and electro-mechanical subsystems like servo and feed positioning drives of the thirty antennas spread over a radial distance about 15 km from the central control building. To control and coordinate new or upgraded subsystems of the telescope, and for performing astronomical observations with enhanced automation, the GMRT M&C system is being upgraded.

In conventional monitor and control systems, as the synchronization and communication protocols are tightly coupled with individual device level controllers, adding new features in response to upgrade requirements requires significant changes and becomes a difficult task. The advent of technologies in micro-controllers like PC104, Rabbit and Labjack, allows running the embedded control software with IP based interfaces which can conveniently handle the different states of the devices, including functionalities like safety and hard real time requirements of the instrument. To exploit these advantages, next generation M&C systems are mainly based on a specifications driven framework and that provides a powerful and flexible way to implement functionalities like coordination between the telescope subsystems, handling their states (like initialization, normal, suspend, restore etc.) and providing a configurable user interface.

A prototype version of a new M&C system for the GMRT, developed in collaboration with the System Research Laboratory of TRDDC-TCS, India [2], uses a generic specifications-driven M&C framework called Sensor Actuator and Control Element (SACE). NCRA has recently built a 15m radio telescope in its Pune Campus for which, a generic, web based M&C system has been developed in collaboration with Persistent Systems Ltd., India [3]. In sections 2 and 3, we describe these M&C systems in detail. In section 4, we note the commonalities between these two systems, and also carry out a comparative study with the M&C systems of other modern telescopes, to highlight the current trends in M&C system development such as: usage of generic frameworks like EPICS, Service Oriented Architecture and SACE. We also discuss the issues relevant to the development of next generation M&C systems for large radio telescopes like the SKA [6].

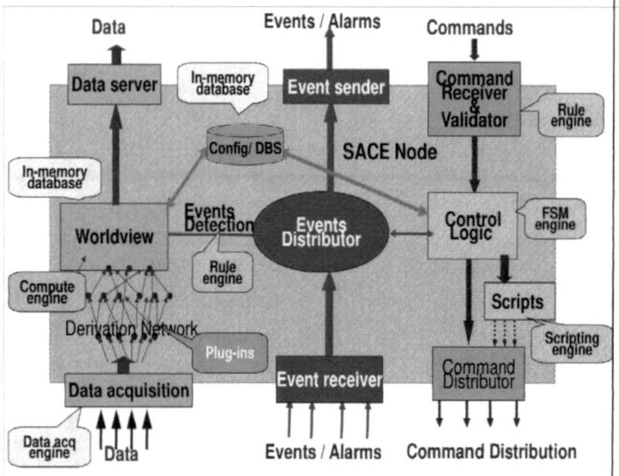

Figure 1 SACE High Level Architecture Diagram

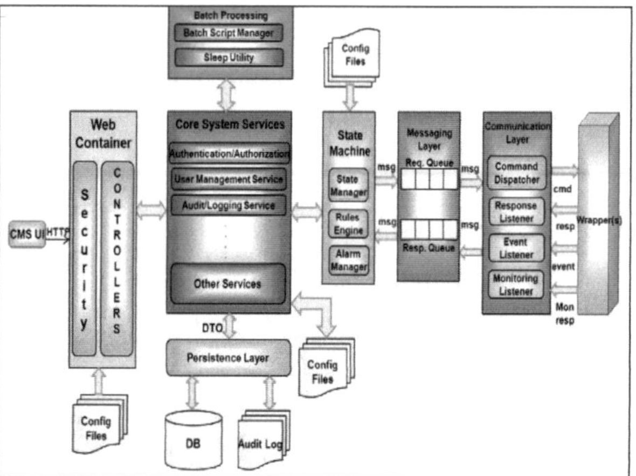

Figure 2. Web based M&C system architecture for the 15m antenna

2. SACE Architecture based GMRT M&C system

2.1. SACE Architecture : The basic building block of the framework is a SACE node (figure 1), that provides command, response and event data streaming interfaces to the *child* and *parent* nodes running locally or remotely on heterogeneous operating system environments. Each node provides standardized M&C functionalities like command-response validation and translation, state machine for control algorithm flow etc. At the application layer, the SACE node functionalities can be configured as required using the input files and rules appropriate to the instrument under control. The core of the SACE is nothing but the Meta model interfaces written in Java which provide wrappers around the components or plug-in modules specified in the initialization file. The SACE architecture partitions the M&C functionalities into components that can be mapped to independent tasks which can be implemented using the available or futuristic alternative technologies. For example, the desired application control logic is implemented using the UniMod java FSM framework, command validation is done using OpenRules which takes inputs in MS Excel format, open source java script engine Rheno is used for the scripting, and Java Swing frame-work is used to auto generate the GUI using XML self-description files. The SACE engine uses *config.ini* file for component selection along with the Java libraries and deployment specifications or configuration XML files, to generate the project specific M&C runtime environment.

2.2. The GMRT M&C development: The SACE node is repeated at each level of the GMRT M&C system, along with specifying the application behaviour using the self-description and FSM logic to perform dedicated tasks. The SACE instance running at any antenna (called *antenna SACE)*, receives and validates the commands from the parent SACE node running at the central building (*CEB SACE*) and distributes it to the individual child *Subsystem SACE* nodes which control analog receiver chain systems and servo and feed positioning systems at the antenna base. Similarly, responses and events gathered from all the *subsystem SACE* are sent to the *CEB SACE* through the *antenna SACE.* A prorotype version of this GMRT M&C system formed by hierarchically composing SACE nodes at each level of M&C, system has been successfully tested on a set of three GMRT antennas.

The SACE architecture has been proven to be based on self-description concepts, best suited for hierarchical composition on multiple devices and provides most of the functionalities required by a M&C system. SACE is a scalable, cost saving and agile solution as it provides interfaces for COTS & customized software.

3. Web based M&C system for the NCRA 15m telescope

The NCRA 15m antenna is a Alt-Azimuth mounted parabolic dish located at Pune. We have developed a web-based M&C system suitable for this antenna and for the 2m Optical telescope of ICUAA [3], based on an architecture which is sufficiently generic to cater to both types of telescopes with most of the differences in specification captured as configuration parameters.

3.1 Web based M&C Architecture and the 15m antenna M&C system implementation:

Figure 2. shows the high level architecture diagram of the M&C system for the 15m antenna. The spring frame-work, used along with the web container, provides runtime environment for the web functionalities like security, concurrency, life-cycle management, transaction and other services. The spring web controllers route the request from the C&M System user interface (CMS-UI) to the Core System Services. The CMS-UI display web-page consists of HTML and Flex components which handle the HTTP requests and responses from the telescope users. The core system services accept and validate the requests from the controller, and pass them to the appropriate service components like batch processing, state-machine and data layer objects. The core system also provides dynamic service level integration and implementation such as user and user-group or role management, catalog management etc. The state-manager is responsible for over-all behaviour of the M&C system and keep updates of live status of all the telescope sub-systems like servo, signal conditioning, sentinel and back-end processing units. Upon raising of alarms or exceeding the threshold levels of telescope parameters, state-machine takes corrective actions to restore the telescope to normal operations. The core-system services & state machine uses XML specifications, self-description and rule files which increase the level of configurability of the M&C system. The communication and messaging layers mainly handle the requests / responses, asynchronous events and monitoring data between the CMS and telescope subsystems wrappers using message queues and socket communication. The wrapper software is generic and configurable, and communicates to all subsystems of the telescopes. The wrapper decodes the XML packets received from the CMS, passes to hardware devices to take action, and after gathering the responses from devices, converts them to XML packets to send to the CMS.

3.2 Salient Features:

Our web based M&C frame-work is a end to end software solution for the radio and optical telescopes M&C system which has the salient features like (i) Context based web-browser interface. (ii) Modular and allow easier hardware/software upgrades. (iii) Scalable and configurable (via xml-DTD specification). (iv) Automatic state-restoration, exception handling using the batch/scripts. (v) Gives alarms notifications through the audio, visual and emails.

4. Current Trends in the M&C System Technologies

Table 1 shows a comparative study of the M&C system frameworks used in the GMRT and the 15m Antenna, along with those for other new M&C systems in upcoming radio telescopes like the Australian SKA pathfinder (ASKAP) [4] and the MeerKAT in South Africa [5]. We note the commonalities like all the M&C systems use Linux OS platform and open source software for the web-browser based UI, Java technologies, programming/scripting languages like Java, C++, python, jsleep and middleware based on the TCP/IP based protocols, *activemq* and *ICE* for the communication. The scalable and specifications-driven framework achieved by using self-description / configuration files in .xml and .ini formats, and state-machine and command / response validation uses rule files. Most of the M&C systems use MySQL for the database.

Table 1: Trend in technologies for the Radio Telescope M&C System

Telescopes	NCRA-GMRT	NCRA 15m Antenna	ASKAP	MeerKAT
Platform	Linux, Solaris	Linux	Linux, MacOSx	Linux MacOSx
Architecture	SACE (Hierarchical)	Spring Framework, Tomcat (Service Oriented)	EPICS (Server-Client Model)	Service Oriented Architecture (SOA, Python based client library)
Programming Languages & Scripting	Java, C++, Mozilla rhino script	Java Technologies, C/C++, jsleep	C++, Java, Python	Python and C++
User Interface	Java Swing based	Web based (java scripts, servlet, flex components	Control system studio, MoniCA server, BOY	Web Interface, Adobe AIR, ipython shell, stallerium soft.
Middleware (Commun.)	JMS, TCP/IP socket lib	Activemq, TCP/IP	Channel Access and ICE	Katcp-python(pypi), TCP/IP
State Machine	UnimodFSM executer	groovi, DRools	IOC-State Notion language	
Database	HSQLDB 1.8, MySQL	HSQLDB (3.2.6), MySQL	Epics Database using PyEpics, MySQL	MySQL
Self-description	.xml and .ini files	.xml and .xls files		.ini and .xml files
Command validation	OpenRules and MS Excel	DRools	EPICS-IOC	Pythonic rules
Alarm Notification and exception handling	Audio/Visual	Audio/Visual/SMPT, JMS2email	CSS BEAST Component – Alarm configTool, JMS2RDB	

5. Summary

Frameworks used for the M&C systems for the GMRT, the NCRA 15m antenna, the ASKAP, and the MeerKAT provide standardized functionality with important features like : (i) Cost-effective proven approach (ii) generic platform (iii) loose coupling (iv) scalable and configurable system (v) automatic state-restoration, exception handling (vi) context-based UI and reporting strategies (vii) Alarms with audio, visual and emails/SMS notification and (viii) integrated access to remote nodes and/or monitoring storage (archives). In conclusion, large resources like money, time, and man power invested during design and develop a customized M&C system and during the up-gradation of the new hardware, can be avoided by using the generic frameworks for M&C systems described here.

Acknowledgement: We thank to the members of telemetry group of GMRT who actively involved in testing of the M&C systems described in this paper. We thank Mr. Amrit Ahuja from the TRDDC, TCS for providing the SACE architecture diagram and Mr. Bhushan Shelke & Mrs. Swati Jadhav from Persistent System Ltd. for making available the web based M&C architecture diagram.

References

[1] Swarup G *et al* 1991 Giant Meterwave Radio Telescope (GMRT) - Scientific objectives and design aspects *Current Science* **60** 95
[2] Chaudhari S, Ahuja A et al. 2009 *Low Frequency Radio Universe ASP conference series* **407** 402
[3] Mohile V *et al* 2009 *Low Frequency Radio Universe ASP conference series* **407** 416
[4] Guzman J C 2009 Preliminary Design of the Australian SKA Pathfinder (ASKAP) Telescope Control System *Proceedings of ICALPCS2009, Kobe, Japan* TUD001 343
[5] Heever L V *et al* 2009 Monitoring and Control MeerKAT South Africa *HPC in observational Astronomy conference, India*
[6] Ahuja A *et al* 2011 *General Assembly and Scientific Symposium : XIIIth URSI* 10.1109/URSIGASS.2011.6051289

IOP Conf. Series: Materials Science and Engineering **44** (2013) 012027 doi:10.1088/1757-899X/44/1/012027

Astrophysical Results of the Mauritius Radio Telescope.

R Somanah[1], N Issur[1] and N Oozeer [2,3]

[1]Department of Physics, University of Mauritius, Reduit, Mauritius

[2] Square Kilometre Array South Africa, Pinelands, Cape Town, South Africa

[3] African Institute of Mathematical Sciences, Muizenberg, Cape Town, South Africa

E-mail: dinesh@uom.ac.mu

Abstract. One of the first scientific justifications of building the Mauritius Radio Telescope (hereafter referred to as MRT) was to complement the Cambridge 6C survey, which is a radio map of most of the northern sky at 150 MHz [1]; the MRT would then be the equivalent of the 6C survey for the southern sky and together we would obtain a whole sky radio map at 150 MHz. When the MRT was built, there were no radio surveys of the southern sky at frequencies less than 408 MHz; the frequency of 150 MHz was also chosen to complement the other radio surveys of the southern sky, which have been done at higher frequencies. Furthermore low radio frequencies like 150 MHz are bound to see new sources that would have been missed at higher frequencies due to the form of their spectra. Interesting features of resolved objects can also be studied in more details. In this paper, a brief description of the MRT will be made as well as the observations and imaging with the MRT data, and some astrophysical results obtained since its commissioning in 1992 (20 years of existence this year 2012).

1. Introduction

Early radio surveys were made at decametre wavelengths, and as higher-frequency receivers became available, observations of higher resolution at centimetre wavelengths became possible. A few major problems encountered by radio astronomers wishing to perform high-resolution, high-sensitivity observations at low frequencies, are the very large aperture dimensions required, phase perturbations caused by the ionosphere [2], and interference from terrestrial transmitters becoming severe.

Surveys are like bread and butter for astronomers as they form the starting point of identification of any class of celestial objects with well-defined criteria. For any survey it is desirable to have nearly uniform sensitivity across the sky. This gives a more reliable overall view of the sky and a more homogeneous sample. Using their statistical properties one can study/infer cosmological evolution. The unbiased and homogeneous samples from surveys are essential for testing models of source evolution. In addition, they provide a list of sources with well defined criteria and define interesting regions of the sky which can be chosen for further studies depending upon the scientific objectives in mind. Most classification schemes for Galactic and Extragalactic objects have emerged from surveys and have henceforth evolved by imaging a given sample of objects from the survey at multi-wavelengths, most often with higher resolution and sensitivity, and also by obtaining polarimetric and spectral line information.

In this paper, we highlight some of the results from the Mauritius Radio Telescope (MRT) as a surveying instrument. In Section 2, we give a brief overview of the MRT. In Section 3, we touch on the imaging problem followed by some of the results obtained using the MRT from galactic to extra-galactic radio astronomy. We also show some of the pulsar work and finally conclude in Section 4.

2. The Mauritius Radio telescope (MRT)

The Mauritius Radio Telescope is a Fourier synthesis array, which has been constructed and operated collaboratively by the University of Mauritius, the Raman Research Institute (India) and the Indian Institute of Astrophysics (India). MRT is situated at Bras d'Eau (Latitude 20.14° South, Longitude 57.73° East) in the North-East (NE) of Mauritius. The array consists of helical antennas which have the best response in the frequency range of between 80 MHz and 200 MHz and could stand cyclonic wind gust above 200 kmh^{-1}. MRT is a T-shaped non-co-planar array with a 2048 m long East-West (EW) arm having 1024 helical antennae and a 880 m long North-South (NS) arm having 15 mobile trolleys with 4 helical antennae each. The helical antennae are mounted with a tilt of 20° towards the south so that they point towards a Declination (δ) of -40.14° at the meridian of transit. This tilt allows a better coverage of the southern sky including the southern-most part of the Galactic plane, a region largely unexplored at meter wavelengths.

The antennae in the EW are divided into 32 groups and there are 16 groups in the NS. The signal from each group is then filtered, down-converted into an intermediate frequency (IF), amplified and sent to the telescope building where it is digitized. The digitized signals are processed in a 1024-channel correlator. Custom written software running on GNU/Linux OS computers is used to transform these correlated signals into raw images called dirty maps.

The MRT uses the technique of aperture synthesis to simulate a 1 km by 1 km filled array. Observations are made with the trolleys in the NS arm at their nearest position from the array centre. The trolleys are then moved further south and the observations repeated 62 times. This process continues until the end of the south arm is reached. The data for each day is added so as to make a 2-D map of the sky. Also, the non-co-planarity of the EW arm combined with the wide primary beam (full width at half maximum (FWHM) $\sim 60°$) have led to new imaging techniques used in CLEANing the raw data. To summarize, wide field imaging with non-coplanar baselines [3] where the data has been collected for more than 5 years, has made the processing of the data into radio images more complex, which have led to many innovative ideas in the imaging process [4, 5, 6, 7].

Although the MRT was primarily designed to conduct the 151.5 MHz survey, it has also been used for pulsar observations. During pulsar observations, only the EW arm is used. The group outputs are added together, with a tracking capability of about 2 degrees for a source transiting at meridian. This corresponds to 8 minutes for an equatorial source. The data is recorded at a fast rate over a bandwidth of 1 MHz. The data processing is done to produce a de-dispersed output in the desired format, including the pulsar profile unique to each pulsar [8].

3. MRT images and astrophysical analysis

3.1. Continuum Survey

The MRT does not produce instantaneous 2-D images of the radio sky seen by the EW and NS groups of helices. It obtains information over a time period from which an image can be constructed later. The imaging is restricted to making 1-D scans along the meridian and the 2-D image is time stacks of these 1-D scans. The 24-h 'dirty' images of the southern sky visible from the MRT site have already been made and the process of deconvolving the images is currently being finalized. It is the first time that galactic and extragalactic sources of the southern sky have been imaged at 150 MHz, with a synthesized beam width of $4' \times 4.6' \sec(\delta + 20.14°)$.

With a very modest resolution, the continuum radio images obtained using the MRT contain a considerable amount of astrophysical information. Fig.1 shows the galactic centre as seen by the MRT. We first extracted a sample of sources with flux densities > 10 Jy at 150 MHz from our images and we have called it the MRT10 sample. Using the spectral index between 150 MHz and 408 MHz [10], we have made another sample of sources with flux densities > 10.9 Jy at 178 MHz (similar to the 3CRR survey [11]) and we have called it the SMRT10 sample. We compared the SMRT10 sample with the 3CRR and found that they have many similarities as expected [12].

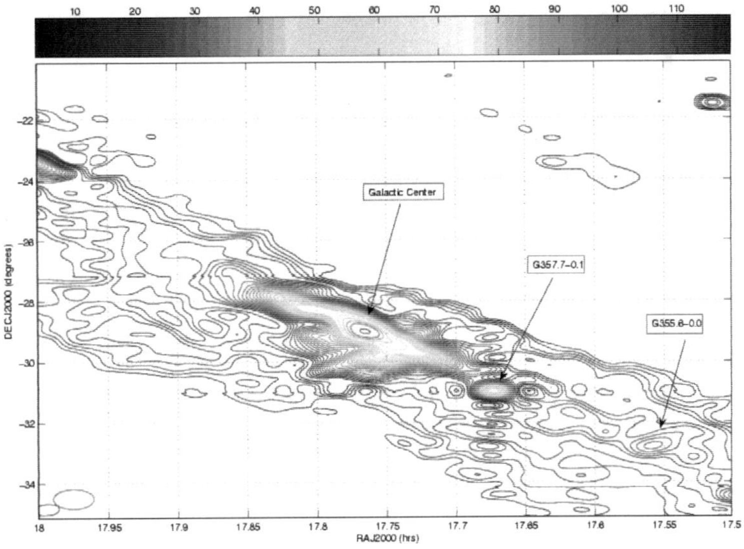

Figure 1. Our galactic plane in the MRT radio images. The contour levels are at: [-0.5, -4, -3, -2, -1, 1, 2, 3, 4, 5, 7, 10, 14, 20, 29, 40, 54, 72, 90, 100, 136, 180, 216, 288] Jy beam^{-1}

3.2. Pulsar observations at the MRT
The MRT has been used to observe some Southern Sky pulsars at 150 MHz. A tracking system allowing observation of a source for $8 \times sec\ \delta$ minutes of time and a fast data acquisition system (FDAS) was added to the MRT. With this new system, pulsars with an average flux density of about 100 mJy could be observed. Sources were tracked between the half power points of the EW group beam.

From observation of the pulsar J0437-4715, we observed a very broad profile with a width at outer half-power points of about 125° in longitude, Fig.2. Three components are observed with a possible suggestion of a fourth component.

4. Conclusions
Until 2010, the MRT was the only radio telescope operating at 150 MHz in the southern hemisphere and the 150 MHz radio images of the southern sky made at the MRT have provided and will provide a wealth of astrophysical data for the community of astronomers. Two radiotelescopes which will also bring new data at low frequencies are the Precision Array to Probe the Epoch of Reionization (PAPER) [13], operating from 120 to 190 MHz and the Murchinson Wide Field Array (MWA) [14], operating from 80 to 300 MHz. Both PAPER and MWA will probe the epoch of re-ionization. Many innovative techniques have gone in the design of the

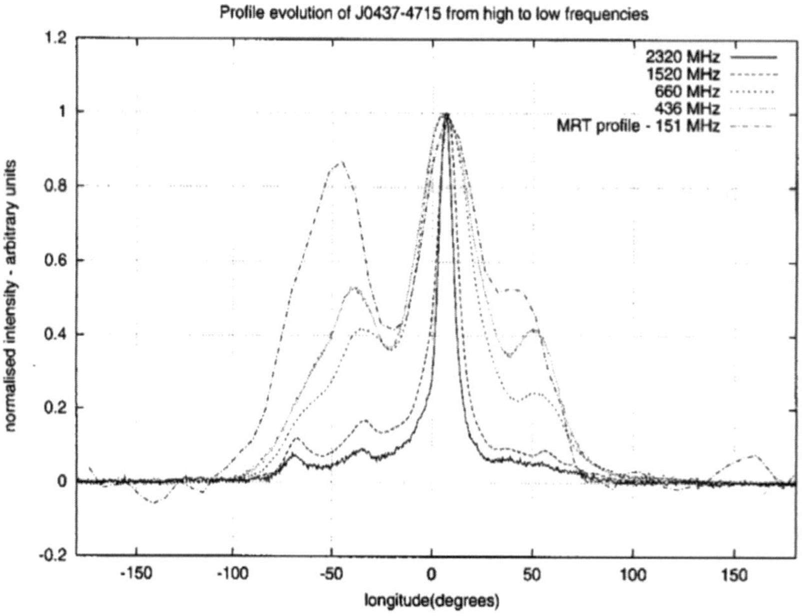

Figure 2. Figure shows the Millisecond pulsar J0437-4715 observed at the MRT, with an average flux density (measured on 27 July 1996) of 620 mJy with a peak of more than 2 Jy, assuming an effective sky background temperature of about 220K.

hardware, software and the mechanical system of the MRT array. Some astrophysical results have been presented in this paper.

The MRT has been used to observe some Southern Sky pulsars at 150 MHz. A tracking system allowing observation of a source for 8 minutes of time and a fast data acquisition system was added to the MRT. With this new system, pulsars with an average flux density of about 100 mJy could be observed.

References

[1] Baldwin J E, Boysen R C, Hales S E G, Jennings J E, Waggett P C, Warner P J and Wilson D M A 1985 *MNRAS* **217** 717
[2] Intema H T, van der Tol S, Cotton W D et al. 2009 *Astron. Astrophys.* **501** 1185
[3] Cornwell T J and Perley R A 1992 *"Radio-interferometric imaging of very large fields: The problem of non-coplanar arrays". Astron. Astrophys.* **261** pp 353-364
[4] Golap K, Issur N H, Somanah R, Dodson R, Modgekar M, Sachdev S, Udaya Shankar N and Sastry Ch V 1995 *Astrophys. Space Sci.*, **228** 373
[5] Golap K, Dodson R, Sachdev S, Udaya Shankar N and Sastry Ch V 1998 *Journal of Astrophysics and Astronomy* **19** pp 35-53
[6] Oozeer N and Udaya Shankar N 2002 *Astrophys. Space Sci.* **282** 43
[7] Cornwell T J, Golap K and Bhatnagar S 2005 ADASS XIV *ASP Conf.* **347**, 86
[8] Issur N H 2002 *Astrophys. Space Sci.* **282** pp 77-85
[9] Somanah R and Udaya Shankar N 2002 *Astrophys. Space Sci.* **282** pp 57-67
[10] Large M I, Mills B Y, Little A G, Crawford D F and Sutton J M 1981*MNRAS* **194** 693.
[11] Laing R A, Riley J M and Longair M S 1983 *MNRAS* **204** 151
[12] Somanah R 2006 PhD thesis *A low resolution survey of the southern sky at 151.5 MHz using the Mauritius Radio Telescope*
[13] Backer D C *et al* 2007 *Bulletin of the American Astronomical Society* **39** p 967
[14] Ord S, Wayth R, Mitchell D and Greenhill L 2007 *Bulletin of the American Astronomical Society* **39** p 744

Institute of Physics
Dirac House, Temple Back
Bristol BS1 6BE UK

CURRAN ASSOCIATES INC
proceedings.com